RAND

California Base Closure

Lessons for DoD's Cleanup Program

David Rubenson, John R. Anderson

Prepared for the
Office of the Secretary of Defense

National Defense Research Institute

The research described in this report was sponsored by the Office of the Secretary of Defense (OSD), under RAND's National Defense Research Institute, a federally funded research and development center supported by the OSD, the Joint Staff, and the defense agencies, Contract No. MDA903-90-C-0004.

Library of Congress Cataloging in Publication Data

Rubenson, David, 1954–
    California base closure : lessons for DoD's cleanup program /
David Rubenson, John R. Anderson
        p.   cm
    "MR-621-OSD."
    "Prepared for the Office of the Secretary of Defense."
    "National Defense Research Institute."
    ISBN 0-8330-2327-6 (alk. paper)
    1. Hazardous waste site remediation—California—Case studies.
    2. Military base closures—California—Case studies.
    3. Hazardous waste site remediation—Government policy—United
States.   4. United States—Armed Forces—Facilities.
    I. Anderson, John R., 1962–   .   II. United States. Office of
the Secretary of Defense.   III. National Defense Research Institute
(RAND Corporation).   IV. Title.
TD1042.C2R83   1995
363.73´84—dc20   95-42191
                                                                              CIP

RAND is a nonprofit institution that helps improve public policy through research and analysis. RAND's publications do not necessarily reflect the opinions or policies of its research sponsors.

Published 1995 by RAND
1700 Main Street, P.O. Box 2138, Santa Monica, CA 90407-2138
RAND URL: http://www.rand.org/
To order RAND documents or to obtain additional information, contact Distribution Services: Telephone: (310) 451-7002; Fax: (310) 451-6915; Internet: order@rand.org

RAND

California Base Closure

Lessons for DoD's Cleanup Program

David Rubenson, John R. Anderson

Prepared for the
Office of the Secretary of Defense

National Defense Research Institute

PREFACE

This report examines the policy innovations implemented to enhance the cleanup of closing military bases in California and considers their application to other aspects of DoD's cleanup program and the national discussion regarding hazardous waste policy. It should be of interest to those involved in federal facility cleanup and hazardous waste policy. The work was conducted for the Deputy Under Secretary of Defense for Environmental Security, within the Acquisition and Technology Policy Center of RAND's National Defense Research Institute, a federally funded research and development center sponsored by the Office of the Secretary of Defense, the Joint Staff, and the defense agencies.

CONTENTS

TABLES

SUMMARY

ACHIEVING MULTIPLE GOALS IN THE DoD CLEANUP PROGRAM

The cumulative effect of public opinion and legislation during the past decade requires the Department of Defense (DoD) to clean up hazardous waste on closing, active, and former DoD sites. Cleanup is primarily governed by the Comprehensive Environmental Response Compensation and Liability Act (CERCLA), the Superfund Amendments and Reauthorization Act (SARA), and the implementing rules contained in the National Contingency Plan (NCP). CERCLA and SARA have been debated intensively and are often criticized for promoting slow, ineffective cleanup. Congress appears to be frustrated and seems determined to reduce annual cleanup funding. This implies that complete cleanup of military bases will be deferred and the viability of the cleanup effort may hinge on setting priorities and achieving meaningful interim goals.

Priorities and interim goals have been discussed extensively in the national debate on cleanup policy. We introduce our analysis, which is presented in the form of an annotated briefing, with a narrative background discussion of CERCLA, SARA, and priority setting. We describe a variety of priority-setting systems, including risk driven, reuse driven, speed driven, jobs driven, and others. Of particular interest is DoD's explicit attempt to implement reuse-driven cleanups on closing bases. This suggests that cleanup should be sequenced according to potential community interest in reusing portions of a base.

Implementation of any priority system depends on the ability to adapt the CERCLA and SARA regulatory framework to that system. We note that many assume that CERCLA and SARA already correspond to a risk-driven priority system and hence require little adaptation for this approach. We then describe the requirements imposed by CERCLA and SARA and highlight the areas in which project managers and/or local regulators have significant flexibility.

Use of this flexibility to implement priority systems and to achieve multiple interim goals is currently being tested on California's closing bases. California has 19 major closing bases, and there is intense economic and political interest in accelerating the cleanup process. More significantly, DoD has an explicit goal of integrating reuse-driven cleanups with risk-driven cleanups and CERCLA and SARA compliance. It has also implemented a number of policy innovations, known as the "fast track," to ensure that local project leaders and regulators take maximum advantage of the flexibility existing in CERCLA and SARA.

Giving significant attention to closing bases in California and examining on-going efforts to achieve multiple priorities within the CERCLA and SARA framework, we identify the following questions, which are the focus of this report:

1. What goals actually govern cleanup projects for California's closing bases? To what extent have cleanup projects been modified to conform to reuse-driven goals?
2. Is risk-driven priority setting a by-product of CERCLA and SARA requirements or is it a distinct priority-setting system?
3. How significantly do reuse-driven priorities differ from risk-driven priorities and/or CERCLA and SARA requirements? What about speed-driven or jobs-driven priorities? Can DoD's goal of harmonizing risk-driven priorities with reuse-driven priorities and CERCLA and SARA requirements be achieved?
4. What is the role of project management in achieving this goal? Do DoD policy innovations help project managers take advantage of the flexibility existing in CERCLA and SARA?

Having defined the policy questions, the remainder of the report is written in briefing form with an emphasis on site-specific concerns. We have chosen this format because the essential trade-offs in an engineering project are easily illustrated in this manner.

Our approach is case-study oriented. We chose this approach because cleanup projects are strongly influenced by site-specific conditions. Aggregate data are easily skewed by anomalous soils and contamination found at a single site. More significantly, we have taken the case-study approach because question 4 involves the role of project management. It is only with a site-specific focus that this variable can be dissected.

GOALS AND PROTOCOLS

We begin the briefing with a review of base closure in California and DoD's goal of achieving multiple cleanup objectives. We discuss risk-driven, reuse-driven, and speed-driven cleanup and the requirements of CERCLA and SARA. We also discuss some of the fast-track policy innovations that DoD has implemented to improve the cleanup process at closing military bases.

We then describe the relationship between CERCLA and SARA requirements and reuse-driven goals. We note that CERCLA and SARA protocols lead to division of a large facility, consisting of many individual sites, into a series of operable units (OUs). Federal Facility Agreements (FFAs) signed by DoD, the U.S. Environmental Protection Agency (USEPA), and the state specify schedules and milestones for the cleanup of each OU. FFAs are enforceable through fines.[1]

Protocols for transferring clean parcels and for dividing a base in a manner that may suggest a reuse-driven strategy are defined by CERFA, the Community Environmental Response Facilitation Act. Although CERFA has few substantive obligations, it does allow for reuse parcels to be identified and transferred if those parcels are uncontaminated. Reuse parcels typically overlay a preexisting division of a base by OUs defined in the CERCLA and SARA Federal Facility Agreement.

Reuse-driven cleanups will be consistent with CERCLA and SARA if there is good correspondence between OUs and parcels. We therefore focus much of our investigation on a comparison of OUs (CERCLA) and reuse parcels (CERFA). We note, however, that reuse parcels, OUs, and FFAs are administrative constructs. Project managers and local regulators can jointly amend each of these. Therefore, simply tracking

---

[1]Six of the 19 major closing bases in California are not on the National Priorities List (NPL) of the nation's worst hazardous waste sites. The lead regulator for NPL sites is USEPA. The lead regulator on non-NPL sites is the California Environmental Protection Agency (Cal-EPA). On non-NPL sites a Federal Facility Site Remediation Agreement (FFSRA) replaces the FFA.

the overlap between parcels and OUs does not in itself provide a complete answer to questions 1 and 2 above. The role of project management is critical (question 4).

TWO CASE STUDIES

Mather Air Force Base (AFB)

Mather AFB is located in the suburbs of Sacramento and has had a cleanup program since 1982. The base was placed on the National Priorities List in 1987 and on the closure list in 1988. As a result, Mather's cleanup started well before CERFA or President Clinton's "Five Point Plan" introduced a reuse emphasis.

Mather is a large base divided into 19 reuse parcels and only three OUs. One OU contains 59 of the 69 contaminated sites at Mather. The cleanup process is based on OUs and involves a nearly basewide effort, with sequential stepping through CERCLA's lengthy investigative protocols. As a result of the complexity and costs associated with thoroughly investigating an entire base, approximately $50 million has been spent on characterizing the site; virtually no remedial action has been taken. On this basis, there is no evidence that the CERCLA process corresponds to either a reuse-driven or a risk-driven approach. Instead, the Mather AFB cleanup appears to be protocol driven.

Current estimates imply that more than $150 million will be expended before cleanup is complete at Mather. Our analysis suggests that by focusing efforts on the most important reuse parcel, which crosses OU boundaries, along with other sites where contamination could spread, a combined reuse- and risk-based strategy can be pursued. Although we cannot argue that total costs to clean the entire base will be reduced, interim goals can be set and achieved at costs well below that required for total base cleanup. This requires moving away from the current designation of Mather as a single cleanup project. There are many low-risk sites that also have limited reuse value. *Interim goals should, in effect, replace the goal of cleaning up the entire base, which is both too problematic and too expensive to provide a realistic policy objective.*

March AFB

March AFB is located near Riverside, California. Half of the base is slated for closure while the other half will continue to operate. The correspondence between reuse parcels and OUs (CERCLA) is poor, with the small number of OUs implying a basewide sequencing of CERCLA protocols just as at Mather.

The cleanup at March cannot be classified as protocol driven because a key DoD policy innovation, the base realignment and closure (BRAC) cleanup team (BCT) consisting of DoD and regulatory personnel working jointly, has adjusted FFA (CERCLA) milestones to facilitate actual cleanup. The BCT is able to do this by taking advantage of DoD's independent authority to conduct removal actions that are forming the basis of the overall project. CERCLA protocols have become largely a formality.

A variety of factors have allowed March AFB to develop this speed-driven approach, most of which involve taking advantage of the flexibility already contained in CERCLA and SARA. One is the creative use of removal actions originally intended for emergencies and not as a basis for long-term cleanup. This has been accomplished by a combination of experienced project managers and regulators who seem to understand the DoD budgeting and contracting system as well as the CERCLA and SARA process. The trust of the community, which may be more difficult to build in more highly charged political climates, has also been an essential factor. The BCT has also developed a competitive system of contractors rather than relying on a single large contractor for the entire facility. This requires more-intensive management on the part of the BCT, but seems to bring greater speed and efficiency. Although cleanup at March is speed driven rather than reuse driven, the approach could be focused on individual parcels to facilitate partial reuse.

OTHER BASES AND CONCLUSIONS

In the last section of the report, we discuss the cleanup process at a number of bases in California, including El Toro Marine Corps Air Station, Tustin Marine Corps Air Station, Mare Island Naval Shipyard, Hunter's Point Naval Shipyard, Fort Ord, the Presidio of San Francisco, and Hamilton Army Airfield. We show that reuse parcels are generally far smaller than OUs and that the latter move projects toward basewide cleanup strategies and away from achieving interim goals.

These cleanups lead us to answer questions 1-4 above in the following manner:

1. CERCLA compliance, not risk reduction, reuse, or speed is still the goal of many cleanups. A reuse goal is hampered by a poor correspondence between OUs and reuse parcels. Under an intense political spotlight, cleanup projects at Fort Ord and Hamilton Army Airfield have at least partially shifted direction toward cleanup and transfer of reuse parcels. However, this level of political attention cannot be the basis for an overall program.

2. CERCLA compliance often leads to protocol-driven cleanups and not necessarily risk-driven cleanup. The meaning of risk-based priorities is not clear after emergency removals have eliminated known exposure pathways.

3. Under the current divisions of the base by OU and parcels, there is often a significant distinction between CERCLA compliance and reuse-driven goals. However, there is no fundamental reason this must occur and no fundamental divergence between a risk-driven approach and a reuse-driven strategy. Multiple goals can be achieved by renegotiating regulatory agreements, redrawing internal base boundaries, and focusing cleanup efforts on the most important reuse parcels and most risky sites. Use of removal actions, an intermediate cleanup step that can be undertaken by DoD without regulatory review, can accelerate this process.

4. DoD policy innovations facilitate use of flexibility existing in CERCLA. The BRAC Cleanup Team concept of teaming DoD project managers and regulators is particularly successful. However, project leaders and local regulators need a better understanding of the CERCLA flexibility that exists and the interim goals the flexibility should be used to achieve.

The answer to the first question implies that DoD's goal of reorienting cleanups toward reuse goals has not been fully met. Reasons for the limited reuse emphasis include project momentum before reuse became a goal and hesitancy to pay the costs of transitioning a project along new engineering directions, existing FFA (CERCLA) obligations, the absence of precisely defined reuse plans, and the need for highly experienced project personnel who can modify complicated technical/legal protocols.

To achieve these multiple objectives for site cleanups we recommend consideration of the following policy ideas:

- Recognize that the total cleanup of many military bases, though desired, is too distant and too expensive to provide realistic policy goals or structure for project management. There should be increased emphasis on identification of realistic interim goals for cleanup and the obstacles to achieving them, rather than on issues aimed at reducing long-run program costs.

- Identify and eliminate obstacles to redrawing internal base boundaries. The USEPA should continue to move away from the policy of fence-line-to-fence-line listing.[2]

- Recognize that while FFAs and FFSRAs strengthen the ability of regulators and communities to influence cleanup projects, they can impose restrictions that prevent efficiency and achievement of interim goals. Alternatives to the FFA and FFSRA process should be formulated and evaluated.

- Realize that for sites remaining in federal hands pending funds for cleanup, the potential for contamination to spread may represent a better basis for establishing risk-based priorities than traditional exposure assessment.

- Note that despite the well-known problems of CERCLA and SARA, it is the experience and dedication of site project management (including regulators), and the extent of support given by

---

[2]We note that this recommendation is consistent with very recent guidance given to EPA regions. See *Enviromental Reporter*, August 18, 1995, p. 773.

higher-level commands, that are the dominant factors in determining failure or success. DoD's investment in human resources for site-level management is inappropriately low given the enormous projects being undertaken.

At a program level at DoD headquarters and/or at the USEPA, we recommend the following to policymakers:

- Review and refine the flexibilities in CERCLA and SARA as summarized in Table 1 and prepare a version of this summary for project leaders and local regulators.
- Provide clearer policy (as opposed to regulatory) guidance than currently exists to encourage the use of removal actions to break administrative logjams.
- Take steps to retain project leaders and regulators who have the experience to "go off line" as they adapt projects toward achieving multiple interim policy goals. The Air Force Base Closure Authority (BCA) should adopt a policy of retaining existing staff when it takes over a base rather than utilizing new BCA personnel.
- Encourage communities to develop more-strategic reuse plans that provide cleanup projects with general guidance for developing reuse-driven cleanups, even when local zoning or planning processes are still in flux.
- Recognize that the DoD project manager has a more complex obligation than simple administrative oversight of large contractors. The project manager must actively engage in project execution, be involved in all engineering decisionmaking, and not allow contractor-led projects to evolve with little policy focus.

At the project management level, we recommend the following to DoD remedial project managers and regulators:

- Review use of the flexibilities contained in CERCLA and SARA.
- Identify interim goals by dividing each base (both active and closing) into sets of sites that are critical for reuse, critical for risk reduction, critical for other policy objectives, and those where cleanup can be delayed. Cleanup projects should then be focused on the most-critical goals.
- Provide greater geographical focus for all environmental programs (munitions removal, lead, asbestos, historic preservation responsibilities, etc.), in addition to cleanup, for the parcels of greatest interest.
- Closely scrutinize potential economies of scale and recognize that many occur only when contemplating basewide cleanup, for which there may not be sufficient funds.
- Similarly, recognize that the administrative economies of scale associated with large remedial contractors may not outweigh the advantages of a competitive contracting structure. Project leaders should tailor the contracting structure to the needs of individual bases and the interim goals for cleanup.

We note that most of these recommendations are relevant for active bases as well as for closing bases.

LIST OF ACRONYMS AND ABBREVIATIONS

AC&W	Aircraft control and wing
AFB	Air Force base
BCA	Base Closure Authority
BCP	Base Realignment and Closure Plan
BCT	BRAC cleanup team
BRAC	Base realignment and closure
Cal-EPA	California Environmental Protection Agency
CERCLA	Comprehensive Environmental Response Compensation and Liability Act
CERFA	Community Environmental Response Facilitation Act
DERTF	Defense Environmental Response Task Force
DoD	Department of Defense
DoE	Department of Energy
EE/CA	Engineering Evaluation/Cost Analysis
FFAs	Federal Facility Agreements
FFSRA	Federal Facility Site Remediation Agreement
FUDs	Formerly used defense sites
GAO	Government Accounting Office
GSA	General Services Administration
IRP	Installation Restoration Program
NCP	National Contingency Plan
NPL	National Priorities List
OB/OD	Ordnance burial/ordnance disposal
OUs	Operable units
PA/SI	Preliminary assessment and site investigation
PCBs	Polychlorinated biphenyls
PCE	Perchlorethylene
RA	Remedial action
RABs	Restoration advisory boards
RI/FS	Remedial investigation and feasibility study
ROD	Record of decision
RPM	Remedial project manager
SACM	Superfund Accelerated Cleanup Model
SARA	Superfund Amendments and Reauthorization Act
TCE	Trichloroethylene
USEPA	United States Environmental Protection Agency
UST	Underground storage tanks
UXO	Unexploded ordnance
VOCs	Volatile organic compounds

1. INTRODUCTION

BACKGROUND

When Congress passed the Comprehensive Environmental Response Compensation and Liability Act (CERCLA) in 1980, it started the nation on a long-term effort to clean up hazardous waste sites. The U.S. Environmental Protection Agency (USEPA) was given responsibility for regulation and enforcement of those sites placed on the National Priorities List (NPL). It was assumed that most polluters were private parties and that CERCLA's "polluter-pays" philosophy would minimize federal expenditures.

Congress amended CERCLA with the Superfund Amendments and Reauthorization Act (SARA) in 1986 and gave additional focus to contamination at Department of Defense (DoD) and Department of Energy (DoE) sites. Consistent with the polluter-pays philosophy, Congress appropriated funds to the DoD and DoE budgets to pay for cleanup mandated by CERCLA and SARA and the associated rules in the National Contingency Plan (NCP).

Since 1986, there has been a growing awareness that federal facilities are among our worst polluters and that liabilities are so large that Congress would, at some point, be unwilling to provide sufficient funds to meet legal and technical requirements. CERCLA and SARA require complete cleanup at each site and establish schedules and milestones without regard to total annual DoD or DoE cleanup budgets. As cleanups move from initial investigations to more-expensive stages of the process, funding requirements expand and are exceeding congressional appropriations.

THE PRIORITY-SETTING DEBATE

The gap between required and available funding has stimulated debate about setting priorities and achieving interim goals in cleanup. Fiscal constraints imply that the CERCLA and SARA goals of complete cleanup, though still desirable to affected communities, may no longer be viable in the foreseeable future.

The following systems, among others, have been discussed for establishing priorities at federal facilities. We focus on priority systems for the sequencing of activities and not those concerned with cleanup standards. Concepts such as *land-use planning* seek to set priorities for cleanup standards based on end-use considerations and do not address sequencing and the problem of annual budget limitations. That type of priority system asks the question, *How clean is clean?* The following systems are more concerned with the question, *What should be cleaned first?*

We note that unless Congress modifies the relevant provisions of CERCLA, any system must meet CERCLA's requirements. These constraints, and the flexibility contained in CERCLA, will be discussed in the following subsection.[1]

Risk Driven

One mechanism for priority setting is to first address sites within a facility that pose the greatest risk to human health or the ecology. This is commonly referred to as a worst-first sequencing. One variation is to place highest priority on those activities that give greatest risk reduction for a given expenditure. In any case, many assume that since the intent of CERCLA and SARA is risk reduction, the law is already compatible with one of these approaches. If true, the priorities and protocols emerging from CERCLA- and SARA-regulated projects would result in a risk-reduction sequencing.

---

[1] Before discussing priority-setting alternatives, we must clarify confusion over the word "site." Typically, DoD or DoE "NPL sites" represent entire facilities composed of numerous smaller contaminated areas. EPA's definition of an entire federal facility as single "NPL site" is known as fence-line-to-fence-line NPL listing. Unfortunately, the individual contaminated areas within a large federal facility are also referred to as "sites." The confusion is more than semantic because the term "NPL site" obscures the possibility of priority setting within a large facility. It also enflames reaction to the debate over priority setting. While for smaller private NPL sites, priority setting may mean choosing which sites not to clean up, for large federal facilities it generally means selecting preferences within a facility.

Unless otherwise noted, we will refer to "sites" as individual areas within a large DoD facility that has been placed on the National Priorities List. The term "NPL site" will refer to the entire federal facility.

One difficulty in validating this assumption is defining risk. Early in a project, institutional controls, such as use of bottled water or fencing off hot spots, will remove all known exposure pathways and contain known risk. In the NCP, risk is given a long-term definition by assuming that a hypothetical site is reused without institutional controls.[2] Based on those assumptions, risk calculations are conducted, and a risk ranking can be developed.

The difficulties with this approach are numerous. First, uncertainties in risk calculations are enormous. Second, assumptions about eventual reuse are speculative given the long time delays before reuse occurs. Third, the approach ignores any definition of risk during the lengthy cleanup project phase, prior to reuse. Finally it gives no extra weight for contaminants that may have the potential to migrate in unexpected ways. In the real world, undetected low levels of contamination may pose more risk than known higher levels, which can be carefully monitored.

Reuse Driven

With the closure of military bases, President Clinton announced a "five point plan" to accelerate cleanup and reuse of closing bases. One of the most critical points was that cleanup of closing bases would be reuse driven.

Reuse planning has now emerged as a distinct priority-setting mechanism. It differs from land-use planning in that it places less emphasis on cleanup standards and more emphasis on the sequencing of cleanup tasks. In reuse planning, nearby communities would identify which sites on a facility are of potential greatest reuse value and hence should be cleaned first. The cleanup of sites of lesser reuse value would be deferred.

---

[2]Institutional controls are measures for reducing risk other than cleaning up a site. They may involve placing a fence around a site, using bottled drinking water, or other measure to reduce exposure without applying a cleanup remedy.

Speed Driven

Frustration with the pace of cleanups has led to the suggestion that annual expenditures be focused on sites where substantial progress is being made. Under this system, cleanup of sites caught in a seemingly endless cycle of studies would be deferred in favor of sites where actual cleanup is being conducted. Although speed alone does not seem to lead to any obvious interim policy objective, the controversy surrounding the pace of cleanup has given this approach political significance.

Jobs Driven, Technology Driven, and Others

It has been suggested, often in connection with base closure cleanups, that cleanup be used to create jobs for communities near the sites. In this priority system, cleanup tasks requiring labor similar to that available in a nearby community would be of highest priority. For example, near Hunter's Point Naval Shipyard, employment needs would best be served by cleanup tasks producing jobs for blue-collar, semi-skilled, and unskilled workers.

Obviously a wide range of other policy goals could also be applied to cleanup priority setting. One suggestion has been to give priority to those projects for which innovative technology is being applied so that long-run costs might ultimately be reduced.

CERCLA AND SARA REQUIREMENTS

Under the current law, any priority system, or combinations of systems, must be consistent with CERCLA and SARA. Section 120 of CERCLA and SARA, and corresponding sections of the NCP, describe responsibilities for federal agencies and establish a sequential process for cleanup:

1. Preliminary assessment (PA)
2. Site investigations (SI)
3. Interagency agreement or Federal Facility Agreement (FFA)[3]

---

[3]Or FFSRA for California-regulated sites. In actuality, CERCLA (Section 120 (e)(1)) calls for the FFA six months after completion of

4. Remedial investigation (RI)

5. Feasibility study (FS)

6. Record of decision (ROD)

7. Remedial action (RA).

This process must be fulfilled at each facility without regard to national budget limitations or obligations at other federal facilities.

Items 1, 2, 4, and 5 involve the sampling, studies, and investigations needed to identify cleanup remedies. An interagency agreement among the USEPA, the lead agency (DoD or DoE), and the state government, sometimes called an Federal Facility Agreement (FFA), is also concluded. The FFA identifies milestones and schedules and becomes the most important regulatory document. It may also specify penalties should DoD fail to meet obligations. The USEPA also retains influence through its authority to approve the record of decision, which specifies final remedies.

While the above list is no more than a logical description of project phases, CERCLA and SARA are controversial because of the detailed protocols specified for each phase, particularly the RI and FS phases. Mandated studies include a human health-risk assessment, a community relations plan, an ecological risk assessment, sampling and analysis plans, health and safety plans, treatability studies, etc. The list of administrative requirements is lengthy and specific. Schedules are specified in the FFA.

This specificity has led some to believe that CERCLA and SARA cannot be easily adapted to achieve interim goals and that meaningful priority setting will require modification of the law. However, a more careful examination indicates that the law is so detailed that there are many contradictory statements, loopholes, exceptions, and alternative approaches. There are also many site-specific considerations that leave substantial discretion to local project managers and regulators. Table 1 highlights five of the most significant areas of discretion and flexibility.

FS. In practical terms this would delay the FFA for lengthy time periods, and most bases have signed the FFA after the PA and SI.

Table 1

Flexibility in CERCLA, SARA, and the NCP

1. The NCP specifies (300.430(a)(1)(A)) that "NPL sites" can be divided into operable units and follow a "phased" approach to cleanup. Phased schedules can be included in FFAs.

2. FFAs can be modified if all signatories agree. Modifications can include changes in schedule as well as internal base boundaries.

3. While FFAs specify schedules, there is significant room for interpretation as to the depth and detail of required studies and administrative reports.

4. Removal actions (intermediate cleanup activities) for urgent situations can be implemented by DoD without regard to the above sequencing. The USEPA retains authority to determine if removals represent final actions in the ROD.

5. The USEPA has substantial discretion in setting cleanup standards. It has a range of both health risk goals (10^{-4} to 10^{-6} lifetime cancer risk to an individual) and land-use scenarios (industrial, commercial, residential) to determine appropriate remedies. The need to incorporate state laws in cleanup standards limits this flexibility.

Of particular importance and often leading to significant confusion is the first point in Table 1. Since an entire DoD facility is viewed as a single "NPL site" in the NCP, this implies that a facility can be divided in different ways to implement phased cleanups. This should create substantial opportunity to divide a large facility into sites corresponding to the different priority systems highlighted above.

BASE CLOSURES AND PRIORITY SYSTEMS

The debate over priority systems has taken place in abstract forums far from the realities of cleanup projects. There has been little consideration of "real-world" conditions, which both limit the ability to implement alternative priority systems and blur the distinction among them. Most important, the proceeding discussion implies that the ability of project managers to utilize fully the tools listed in Table 1 will be decisive in determining if CERCLA and SARA can be adapted to accommodate any priority system.

DoD's policy goal of reuse-driven cleanups for closing bases gives us an opportunity to link this abstract discussion with ongoing projects. Compliance with CERCLA and SARA and reuse-driven cleanup are goals of the policy. In response, some have accused DoD of abandoning a more traditional risk-driven approach in favor of economic concerns. Those making such accusations assume that CERCLA and SARA are already consistent with a risk-driven approach.

DoD leadership has maintained that reuse, risk, and CERCLA and SARA compliance can be achieved simultaneously and has developed a number of policy innovations to help achieve these multiple goals. Ongoing base closure cleanup projects give us the opportunity to test the utility of these innovations, the feasibility of achieving multiple goals, and the role of project management in the process.

GOALS OF THE BRIEFING

The following annotated briefing summarizes our examination of several cleanup projects at California's closing bases. Our purpose was to monitor the implementation of reuse-driven cleanup projects and to bring a site-level orientation to the issues highlighted above. Specifically, we monitored several cleanup projects to answer the following questions:

1. What goals actually govern cleanup projects on California's closing bases? To what extent have cleanup projects been modified to conform to reuse-driven goals?
2. Is risk-driven priority setting a by-product of CERCLA and SARA requirements or is it a distinct priority-setting system?
3. How significantly do reuse-driven priorities differ from risk-driven priorities and/or CERCLA and SARA requirements? What about speed-driven or jobs-driven priorities? Can DoD's goals of harmonizing risk-driven priorities with reuse-driven priorities and CERCLA and SARA requirements be achieved?
4. What is the role of project management in achieving this goal? Do DoD policy innovations help project managers take advantage of the flexibility existing in CERCLA and SARA?

The briefing describes several DoD facilities and our overall investigation included several more. Much of the briefing focuses on lessons from the cleanup of March and Mather Air Force Bases (AFBs). The more detailed discussion of these two projects allows for a general description of cleanup projects and the distinctions between alternative approaches to cleanup.   After this structure is established, we then discuss the cleanup of several other bases in terms of the above questions.

2. THE BRIEFING

*California Base Closure Cleanup:
Lessons for DoD's Cleanup Program*

0195PR-##

Figure 1

This document reproduces the charts, text, and supplementary material for a RAND briefing on cleanups at closing and realigning bases in California. As indicated by the title, the emphasis of the briefing is to extract lessons from the base realignment and closure (BRAC) process that will improve hazardous waste cleanup at other closing bases as well as cleanup on active bases and formerly used defense sites (FUDs).

The focus on California's bases is purposeful. The economic effect of base closures and realignment has produced a strong political consensus to accelerate the notoriously slow cleanup process. At California's closing bases, DoD officials, federal regulators, and state regulators are attempting to move forward with greater cooperation and speed. There is a purposeful effort to blend reuse-driven priorities with "risk reduction" and compliance with regulations. A number of policy innovations have been introduced to provide local project leaders with the incentives to take advantage of existing flexibility to achieve these multiple goals.

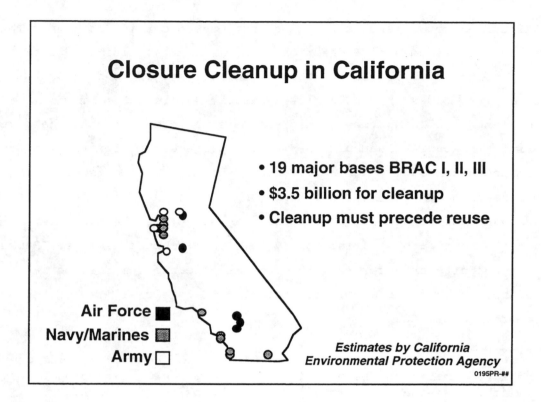

Figure 2

Figure 2 provides an overview of the closing bases in California. Nineteen major bases and many smaller bases have been named in BRAC I, II, and III rounds. At the time of this report, the Long Beach Naval Shipyard and McClellan AFB have been proposed for the 1995 BRAC round, though the final list must be approved by the president and Congress. Of the 19 major bases, 13 are on the NPL of the nation's worst hazardous waste sites.[1]

_____

[1]We note that NPL listing may not indicate that one base is more contaminated than another. DoD, USEPA, and the California Environmental Protection Agency (Cal-EPA) have engaged in a long, well-publicized discussion about whether to list additional bases in California. Flexibility, effect on cleanup speed and cost, and the controlling agency--not degree of contamination--have been the key issues. A base like Mare Island Naval Shipyard is clearly one of the most polluted bases in California, but it is not on the NPL. Mare Island's base closure plan highlights vast sources of diverse types of contamination. The USEPA recently chose to list the Concord Naval Weapons Station, an active base, largely because of its location on the Bay/Delta and the new NPL scoring system, which gives high priority to ecological risk. However, Mare Island is also on the bay and has not been listed.

The scope of cleanup activities that must occur is enormous. The California Environmental Protection Agency (Cal-EPA) estimates total cleanup costs at $3.5 billion, though the estimate is crude and has risen several times in the last few years.[2] Should McClellan be added to the list of closing bases, this estimate would skyrocket. The cost is critical because cleanup is normally a prerequisite for reuse of the base. Reuse can occur without cleanup if DoD chooses to issue a long-term lease, though recent court decisions make this option less certain. The issuance of sequential short-term leases may also be possible, though that approach will inevitably lead to a court challenge. Reuse through property transfer requires that all cleanup remedies be in place, and there is no assurance that funding levels will match the scale of requirements.

---

[2]A recent Government Accounting Office (GAO) report suggests there may be a systematic underestimation of cleanup costs on closing bases. See *Military Bases Environmental Impact at Closing Installations*, GAO/NSIAD-95-70, 1995. The $2.5 billion estimate is based on an internal memo from Cal-EPA dated September 1993.

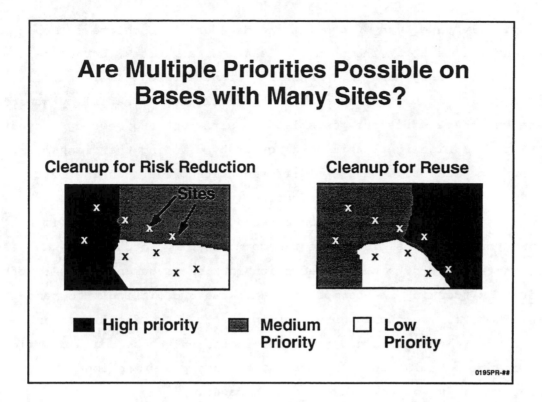

Figure 3

Figure 3 portrays a key characteristic of most closing military bases that was discussed in the introduction. Although listed as a single NPL site, a large military base is typically composed of dozens of contaminated areas that are also referred to as sites. When a single site is discovered that motivates NPL listing, the entire base has normally been placed on the NPL.[3] The practice, which was USEPA policy and is slowly being modified, is commonly known as "fence-line-to-fence-line listing."

Multiple sites imply that interim goals for cleanup may be an important consideration. As indicated by costs shown in Figure 2, the average major California military base cleanup costs over $100 million

---

[3]If the facility is not listed on the NPL, primary CERCLA and SARA authority is transferred to the state, which also treats a facility in a fence-line-to-fence-line manner. We also note that EPA has just recently begun to formally abandon the "fence-line-to-fence-line" approach.

and typically takes many years, if not decades, to complete.[4] Broad agreement exists about the need to clean military bases, but multiple sites at a military base imply that different strategies with different interim goals can be used to achieve total cleanup. The policy question is whether these different goals lead to different engineering projects.

Figure 3 highlights this distinction by dividing the cleanup tasks on a hypothetical base into two different categories. On the left, cleanup is sequenced for risk reduction (to human health or the ecology), on the right for economic reuse. The sharp contrast is not uncommon because reuse may be best accommodated by focusing on the least contaminated part of a base. Reduction in risk to human health is often achieved by focusing on the most contaminated and difficult to remediate sites. However, we should note that it is often the facilities and infrastructure that motivate reuse interests, and these could overlie significant contamination. Therefore, the sharp distinction in Figure 3 is representative of some, but not all, bases.

---

[4]Estimates for cleanup costs of each of the 19 major closing bases are quite crude. The large naval shipyards, Hunter's Point and Mare Island, have estimates approaching $300-$400 million for cleanup. A few smaller bases, such as the Salton Sea Naval Weapons Test Center, the San Diego Naval Training Center, and Treasure Island are relatively small cleanups in the range of $10-$20 million. More than half of the 19 bases are estimated between $100-$200 million. All cost estimates are proving to be significantly low as new contamination is found and additional engineering detail for remedies is developed.

**Can Budget-Limited Cleanups
Blend Multiple Concerns?**

- **<u>Risk</u> reduction is the assumed "<u>legal</u>" goal**
    – Are they the same?
- **<u>Reuse</u> is the Clinton policy goal**
- **<u>Speed</u> is a generally desired goal**
- **Can we achieve them all within:**
    – the current legal framework?
    – shrinking budgets?

0195PR-##

Figure 4

The distinctions highlighted in Figure 3 point to one of the central issues confronting California base closure cleanups. This challenge is discussed in Figure 4.

As noted, the goal of the CERCLA and SARA regulatory system is to reduce the risk posed to human health and the environment. This traditional risk-driven goal differs from the president's announced goal of reuse-driven cleanups. In addition, the notoriously slow pace of cleanup has made "speed" an important policy goal. Regulated communities and regulators are anxious to point to examples in which rapid cleanup is occurring.

DoD has embarked on an ambitious program to simultaneously achieve risk-driven, reuse-driven, and speed-driven cleanups on California's closing bases. Some have criticized the multiple objectives, arguing that human health risks will be neglected if reuse considerations govern

the cleanup process.[5] Others argue that a focus on speeding cleanup
will reduce the care and thoroughness required to reduce risk at a site.
Since CERCLA and SARA are often assumed to represent a risk-reduction
approach, some assume that an emphasis on reuse or speed may be
inconsistent with legal requirements. If so, then legally implementing
reuse-driven or speed-driven priorities would require using the
flexibilities in CERCLA and SARA (Table 1) to adapt the regulatory
framework to be consistent with these aims.

We also emphasize that while the goal of CERCLA and SARA is risk
reduction, it is only an assumption that these laws lead to engineering
projects governed by risk-driven priorities. If this assumption proves
invalid, then the "legal" goal of CERCLA compliance may represent a
fourth competing objective, which, because of the power of regulatory
enforcement, will take precedence over risk, reuse, or speed.

---

[5]At the January 1994 meeting of the Defense Environmental Response
Task Force (DERTF), a congressionally mandated review committee, at
least two task force members expressed doubts that reuse-driven cleanups
could adequately protect human health.

Outline

⟶ • **Communities, governance, and goals**
- **Mather AFB ("legal," risk, reuse)**
- **March AFB (speed)**
- **Other bases and conclusions**

0195PR-##

Figure 5

Figure 5 illustrates the outline of this briefing. In the following pages, we give more-precise meaning to alternative goals for cleanup and the legal framework that governs cleanup projects. We also discuss the political pressures that inhibit project leaders from utilizing innovative approaches and DoD's efforts to overcome these factors through its "fast-track" policy. We then formulate policy questions to be explored at the site level. These questions involve both the status of current projects and issues related to the achievement of multiple goals.

Following that, we turn to case studies of Mather and March Air Force bases. The former highlights the distinction among cleanup strategies, emphasizing risk reduction, legal compliance, and reuse. The latter provides an example of how to increase cleanup speed. We then discuss several other cleanup projects and draw conclusions about the compatibility of alternative cleanup policy goals.

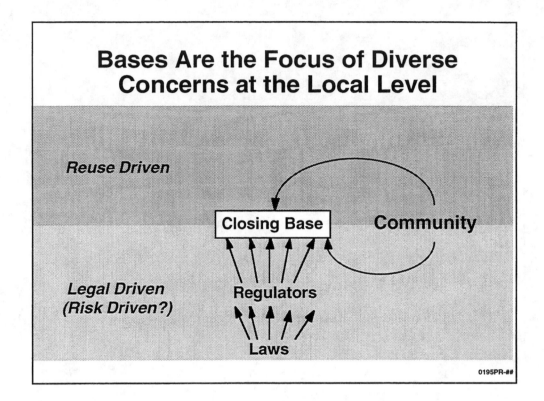

Figure 6

Figure 6 portrays the complex interests that are brought to bear on closing bases by communities surrounding these bases. Many feel these pressures inhibit cleanup project managers from utilizing existing flexibility (Table 1) and other innovations to achieve multiple goals. These political and regulatory complexities may be encouraging an overly cautious approach.

Generally, competing community interests divide along the "reuse" and "risk-reduction" goals that DoD is seeking to address simultaneously. Chambers of commerce, real estate interests, and other civic groups may place a high priority on redevelopment of the site. For such groups, an abandoned military facility could represent a long-term symbol of community decay and failed revitalization. Conversely, the facilities and infrastructure may represent economic values that can stimulate recovery.

Although reuse interests may, to a varied extent, overlap with risk reduction, individuals living near the boundaries of a base, and those with environmental concerns, may have a singular focus on risk reduction. As illustrated in Figure 3, a reuse emphasis may lead project managers toward attacking the easiest hazardous-waste problems first. A risk-reduction emphasis might imply a worst-first strategy.

Those primarily concerned about risk reduction have taken comfort in their assumption that the traditional legal structure supports this goal. This assumption and the uncertainty surrounding its validity is illustrated by the collocation of the "legal" and risk-driven goals in Figure 6. The multiple arrows in Figure 6 reflect the multiple laws and regulators that would be responsible for ensuring this focus. In addition to complying with CERCLA, cleanup must comply with numerous environmental statutes enforced by numerous regulatory bodies. Issues such as the effect of cleanup on wetlands or endangered species require the engagement of other federal regulators and could involve distinctly different approaches than those involved in hazardous waste cleanup and CERCLA.

State regulators are the lead regulators at non-NPL sites and also have numerous ways to engage in an NPL cleanup, including claiming natural-resource concerns and invoking state hazardous-waste law, clean-water law, etc. State cooperation is an essential element in any cleanup project. This is especially true in California where numerous state laws and regulatory bodies can bring their authority to bear.

This complex legal and political framework will inevitably bring a significant degree of caution and conservatism to DoD project managers and regulators. At some bases there is extremely detailed regulatory review and intense community scrutiny for virtually every step of the process. This scrutiny is partially a result of the not altogether positive historic relationship between community groups and DoD facilities. This relationship has begun to change during the past few years, but distrust remains, leading many projects toward literal regulatory compliance, sometimes at the expense of substantive progress. Some community groups and regulatory agencies may be hesitant to give DoD the flexibility to pursue multiple objectives.

Figure 6 portrays a complicated legal and regulatory framework. Diverse community interests may try to influence the cleanup process by addressing local congressional representatives and state legislators, seeking to air grievances in the media, or in any number of other ways. CERCLA cleanups are perceived as producing few cleanup results and being overly politicized and overregulated. The addition of powerful economic interests with stakes in reuse further complicates an already complex situation.

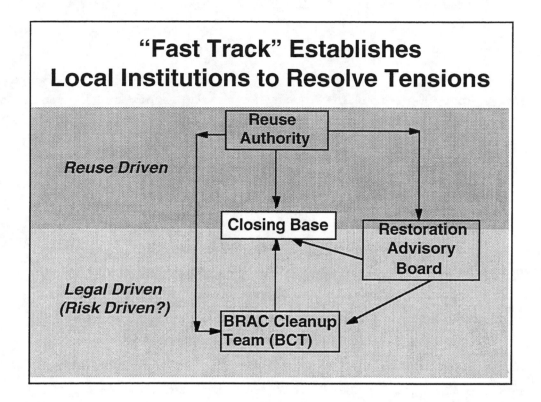

Figure 7

The negative effect of the regulatory and political climate on innovation has stimulated the introduction of several policy innovations aimed at better organization of these complex forces and reduction of the disincentives for project managers and local regulators. Figure 7 highlights some of the key innovations that have become known as the "fast-track" policy.

The key element of the fast track is an attempt to organize and institutionalize the widely diverse interests that could either intentionally or inadvertently impede flexibility and innovation. By recognizing and formalizing these interests, DoD hoped to build a tradition of cooperation and common goals that would overcome historical rivalries and achieve the multiple goals highlighted in Figure 4.

BRAC CLEANUP TEAM

Perhaps the most critical element of the fast track is the BRAC Cleanup Team (BCT). Rather than pit the DoD cleanup project teams against an array of diverse regulators, the BCT concept aims to pull regulators and DoD project leaders toward a common goal. In the BCT concept, Cal-EPA and USEPA assign a single point of contact for each base to work directly with the DoD remedial project manager (RPM). These regulatory representatives are, in theory, empowered to speak for all regulatory bodies while working directly with the DoD project leader. The intent of the policy is to facilitate and sharpen decisionmaking, reduce the number of regulatory interfaces, and give both regulators and DoD project managers a deeper understanding of each other's concerns.

In our judgment, the BCT has dramatically changed institutional and personal relationships. Review of documents has been accelerated and regulators are often exposed to problems of project implementation for the first time. Given the numerous reviews and close regulatory scrutiny associated with the CERCLA process, the BCT serves an invaluable role.

Despite its appeal, the BCT has not completely fulfilled its intended vision. Cal-EPA's goal of centralizing all state regulatory concerns within a single entity has not been achieved. The state Fish and Game Department is completely outside the jurisdiction of Cal-EPA, and Cal-EPA itself consists of highly autonomous boards. One example is the State Water Board, which is both autonomous and decentralized. It is not uncommon to find a water board representative acting as a full additional participant on the BCT alongside the traditional Cal-EPA representative.[6]

We note that building a tradition of cooperation may not in itself be sufficient. Existing traditions for cleanup projects have evolved over many years in response to the climate of distrust. The next step is to use the newfound cooperation to facilitate the achievement of multiple objectives.

_____

[6]Department of Toxic Substances Control is the lead Cal-EPA agency on base closure and is responsible for assigning the state BCT member.

RESTORATION ADVISORY BOARDS AND REUSE AUTHORITIES

Equally important as the regulatory climate is the attitude of local communities toward cleanup. DoD has developed restoration advisory boards (RABs) in an effort to ensure a consistent level of community involvement and partnership with DoD personnel. The boards are jointly chaired by the DoD installation representative and a community leader elected by the community members of the board.

The RABs generally meet on a quarterly basis (or more frequently if desired by the board members). Generally any community member who has desired to be part of the RAB has been allowed to do so, though there have been controversies and some exclusions, particularly when members living considerable distances from the site seek to gain membership. At a typical meeting, DoD project leaders, contractors, and other members of the BCT deliver presentations about the project, answer questions, and take recommendations from the board. The focus varies from base to base, as does the level of community engagement.

The role of the RAB is advisory. As indicated in Figure 7, its major challenge is to ensure that the approach to risk reduction is consistent with community concerns. RABs will also need to ensure that reuse-driven cleanups are acceptable from a risk-reduction perspective. At some bases, community members have joined the RAB because of their interest in reuse, only to be disappointed because responsibility for reuse planning resides with the reuse authority.

Reuse authorities are charged with arriving at a reuse plan for the closing installation. They vary in degree of formal authority, requirements for membership, and ability to speak for the community. In some cases, like Fort Ord, the reuse authority has been legally defined by state law.[7] In other cases, like George AFB and MCAS El Toro, reuse plans have been contentious, with different parts of the community having divergent visions.

The activities of the reuse authority are important because the community reuse plan provides the strategy for a reuse-driven cleanup. The inability to arrive at a widely accepted reuse plan could dissuade

---

[7]Fort Ord Reuse Authority Act, Section 67650, California Code.

DoD project managers and regulators from making difficult adaptations in order to reflect reuse interests.

CHALLENGES

The fast-track institutions, and in particular the BCT, have the challenge of synthesizing diverse community and regulatory interests into a focused cleanup project. More specifically, the BCT has the task of utilizing the flexibility in Table 1 to adapt the CERCLA framework to "fit" diverse policy goals of the cleanup project. Given the goals stated in Figure 4, at a minimum this involves ensuring consistency between division of the base by CERCLA and SARA (item 1, Table 1) with divisions of the base relevant to reuse. It also means adapting both these divisions to ensure that community concerns about risk are also addressed, and overcoming those aspects of CERCLA and SARA that inhibit project speed.

Operable Units (1980), Parcels (1992)

Operable Units
Legal (Risk?)

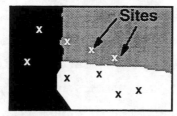

Sites

Parcels
Reuse

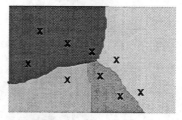

• **Federal Facility Agreements (FFA)**
• **Schedules/penalties**

• **CERFA**
• **Clean parcel transfer**

0195PR-##

Figure 8

Figure 8 provides a more graphic illustration of the challenges facing the fast-track institutions in regard to harmonizing reuse-driven cleanup with CERCLA and SARA compliance. As shown on the left, the legal structure, which is often assumed to be synonymous with a risk-driven priority system, is embodied by the 1980 CERCLA legislation. Under CERCLA and SARA, federal facilities sign FFAs with the USEPA and state officials.[8] Most FFAs divide a base into a series of operable units (OUs) and specify schedules, cleanup milestones for OUs, and penalties if cleanup targets are not achieved. FFAs are enforceable.

Reuse considerations must be integrated into Community Environmental Response Facilitation Act (CERFA) protocols, which became law in 1992, 12 years after CERCLA was enacted. Under CERFA, the base is divided into parcels. Requirements for a parcel to be "CERFA clean" are quite stringent, with the burden of proof to show that a parcel is

---

[8]For non-NPL sites in California, a Federal Facility Site Remediation Agreement (FFSRA) replaces the FFA.

clean rather than just being free of known contamination. As an example of the stringency of regulatory interpretation, 121,200 acres at BRAC I and BRAC II bases were declared "CERFA-clean" by DoD, but regulators concurred with this designation for only 34,499 acres.[9]

An obvious issue is the compatibility of OUs and CERFA parcels. With OUs generally defined prior to CERFA, there is no obvious reason to expect a significant degree of overlap. This would make it difficult to focus cleanup efforts on a parcel because such activities would involve multiple OUs and be inconsistent with the approach embedded in the FFAs. CERCLA contains far stronger and more specific regulatory obligations and will inevitably take priority over CERFA. We do, however, note that the BCT can modify FFAs and redraw OU boundaries to conform to a reuse-driven strategy. However if OU boundaries were conceived to correspond to a risk-driven strategy, then such adjustment may not be desirable.

In many ways CERFA is aimed at reversing the effects of fence-line-to-fence-line listing of a military base. This listing policy was adapted from the regulation of private sector hazardous-waste sites and based on the notion that federal facilities needed to be treated in an identical manner. However, DoD and DoE facilities tend to be much larger than private sector sites and often contain vast tracts of land far from sources of contamination. Under fence-line-to-fence-line rules such tracts have to be proved clean by strict criteria in CERFA. In an identical physical situation with multiple private owners, clean land not adjacent to the contamination, or not held by the owner of the contaminated site, would never be included on the NPL. *Thus, in treating federal facilities identically with those in the private sector, the law has inadvertently treated them quite differently.*

---

[9]GAO, *Military Bases: Environmental Impact at Closing Installations*, GAO/NSAID-95-70, Washington, D.C., February 1995.

Site-Specific Issues

- **What interim goals actually govern cleanups?**
- **Are legal protocols risk driven?**
- **Do OUs ("legal") correspond to parcels (reuse)?**
- **How can project management help achieve multiple goals?**
  - **How do DoD policy innovations support project managers?**

0195PR-##

Figure 9

Figure 9 restates the questions posed in the introduction and places them in the context of our examination of actual projects on California's closing bases.

Our first concern is the actual status of projects relative to the goals discussed in the preceding pages. Closely associated is the extent to which CERCLA and SARA protocols--largely based on OU definitions--correspond to a risk-reduction strategy. Given the difficulty in defining risk discussed in the introduction, this requires examining risk from several perspectives. Of independent importance is the relationship between reuse parcels and OUs. Under limited budgets, the feasibility of implementing a reuse-driven cleanup may be dependent on the ability to focus resources on selected parcels. If such parcels cross several OUs, such focus may be impossible to obtain.

Finally, and perhaps most important, our site-specific examination seeks to identify the role of project management in obtaining multiple goals. Since parcels, OUs, and FFAs are administrative constructs that can be amended, there is no reason why they should inhibit changes in cleanup strategy.   Project management has the responsibility of ensuring that these constructs are not obstacles.   Fast-track innovations are intended to ensure that the political and regulatory complexity does not inhibit project managers from making the necessary adaptations.

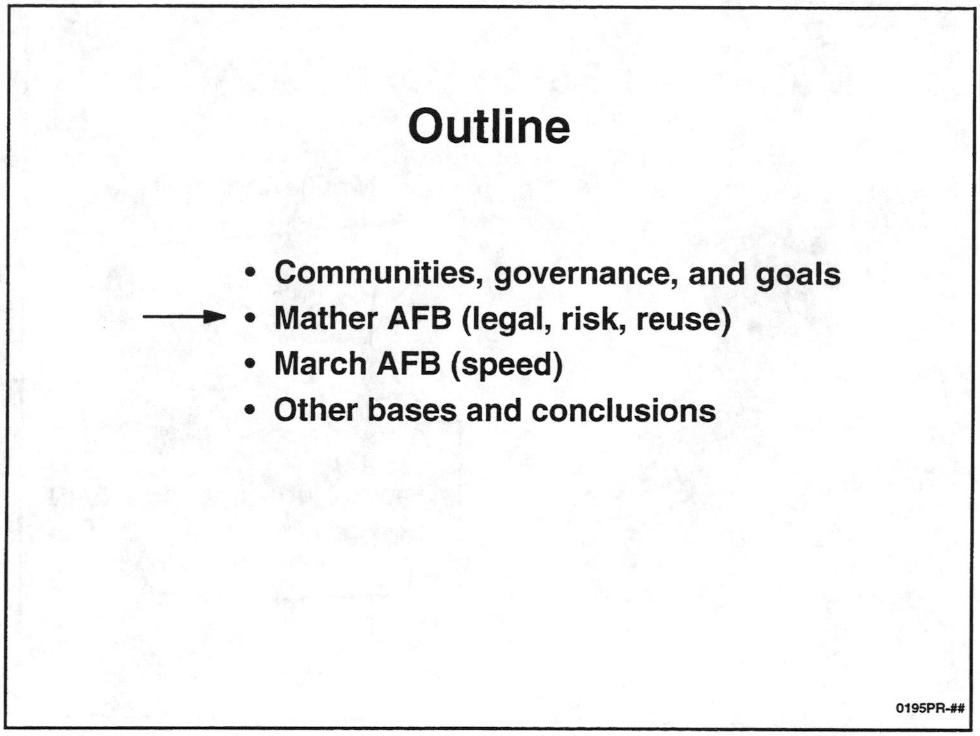

Outline

- **Communities, governance, and goals**
→ - **Mather AFB (legal, risk, reuse)**
- **March AFB (speed)**
- **Other bases and conclusions**

0195PR-##

Figure 10

In the following pages, we consider two examples that provide insight into the questions highlighted in Figure 9. First, we turn to Mather AFB, located outside of Sacramento, to highlight the distinctions among cleanup strategies oriented toward reuse, risk reduction, and legal compliance. Then we examine March AFB, which illustrates some of the constraints on, and opportunities for, speeding the cleanup process.

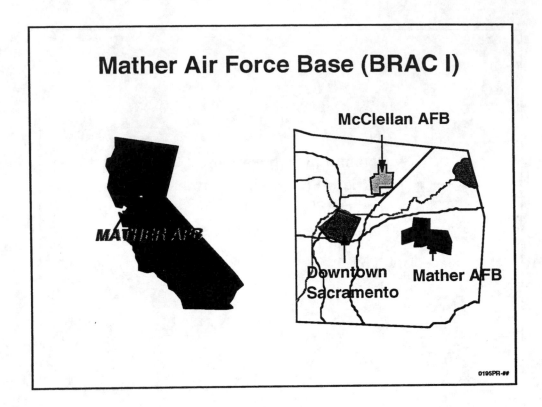

Figure 11

Figure 11 highlights the geographical location of Mather AFB. Located outside Sacramento, the area has displayed intense economic growth over the last decade. However, interest in base reuse has been dampened by California's recent recession. Community concerns about health risk from Mather may also be somewhat reduced because of the intense community interest in McClellan AFB, which is now slated for closure.

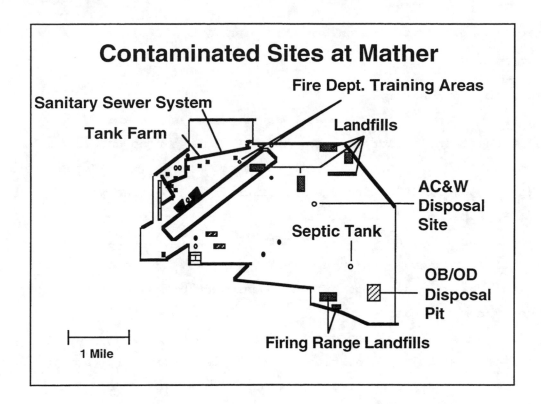

Figure 12

CONTAMINATION AT MATHER[10]

Figure 12 illustrates the overall cleanup challenge at Mather AFB by depicting many of the 69 sites that have been identified in DoD's Installation Restoration Program (IRP). Characteristics of the IRP sites at Mather are typical of sites at many military bases on the NPL. Former uses at Mather include burn pits, a sanitary sewer system, ditches, landfills, underground storage tanks, disposal sites, fueling aprons, maintenance yards, etc. Contaminants found at these sites include solvents (i.e., volatile organic compounds (VOCs)), petroleum

---

[10] Information about Mather AFB was collected from various documents produced by Air Force personnel and contractors at Mather AFB. Documents included Mather's *BRAC Cleanup Plan* and the *Environmental Baseline Survey*, among others. Other sources of information provided during various meetings and conversations include various personnel from Mather AFB, regulators from USEPA and Cal-EPA, representatives from the County of Sacramento, Air Force contractors for Mather AFB, and members of the RAB.

and related products, polychlorinated biphenyls (PCBs), metals and munitions waste, etc.

Threats to Groundwater

Based on characteristics such as former uses, hazardous constituents, and impacted media, the IRP sites can be categorized according to their potential to threaten groundwater. At Mather, sites threatening groundwater include the aircraft control and wing (AC&W) site (IRP site No. 12), which has contributed to a defined plume of contamination, and one of the perimeter landfills (IRP site No. 3) and a former fire training area/burn pit (IRP site No. 8), which are suspected of being the sources of one or two groundwater plumes extending north and west from the base.

In each case involving groundwater contamination, and fairly typical of groundwater contamination problems in general, the constituents of concern (i.e., the contaminants) are VOCs. These compounds, including, for example, trichloroethylene (TCE) and perchloroethylene (PCE), are extremely mobile in soil and water, are volatile (i.e., when exposed to air, they quickly spread in gaseous form), and are considered human carcinogens. Based on these factors, other IRP sites at Mather that involve solvents may be considered to present significant risk to local groundwater resources and to individuals who might be exposed if soil containing these substances were disturbed. Such sites at Mather include the sanitary sewer system (IRP site No. 23), an underground storage tank (UST) tank farm (IRP site No. 39), and a former septic tank (IRP site No. 17).

Soil Contamination

Other sites at Mather involve soil contamination but may pose a less-direct threat to groundwater. One of the most serious of these with regard to a potentially significant risk to human health is the ordnance burial and ordnance disposal (OB/OD) site (IRP site No. 69). This site was formerly used to dispose of munitions waste. Residual contaminants such as dioxins and furans remain at the site and could present a significant health hazard to individuals who come into contact with these compounds.

The remaining 62 IRP sites at Mather consist primarily of former UST sites, landfills, disposal sites, maintenance facilities, fueling aprons, and various fuel spill sites. In general, contaminants associated with these sites are petroleum and related compounds (e.g., petroleum hydrocarbons, benzene, toluene, xylenes), or materials such as asphalt and domestic solid waste. These compounds and materials present a much less significant threat to local groundwater resources and proximal soil when compared with VOCs because of lower mobility and volatility. However, they could pose a risk if individuals were to come in direct contact with contaminated soils.

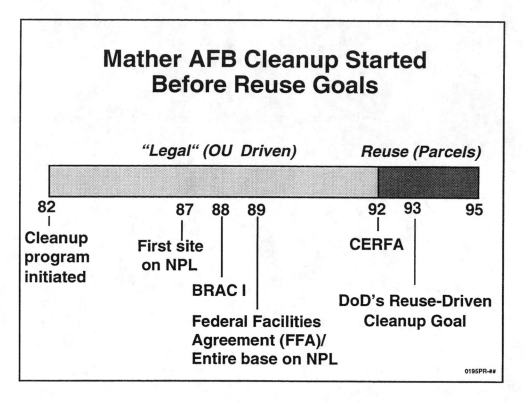

Figure 13

Figure 13 provides a brief historical overview of the cleanup project at Mather AFB. The project began in 1982, well before Mather was placed on the NPL. In 1987, the first site was placed on the NPL. By 1989 the USEPA, in keeping with its general policy of fence-line-to-fence-line listing, had placed the entire base on the NPL. An FFA was signed in 1989. The base was placed on the BRAC I list in 1988 and is now closed. A skeleton staff of the Air Force Base Closure Agency manages the property and the cleanup operations.

Figure 13 highlights intended changes in cleanup priorities resulting from policies initiated in 1992. Cleanup efforts had been on-going under a priority system established by CERCLA and SARA. In 1993, President Clinton's five point plan gave new political emphasis to reuse but did not alter the mechanisms by which legal compliance is achieved.[11] Adoption of the president's goals required not only the

---

[11]President Clinton's five point plan for base closure includes (1) grants to affected communities, (2) a single federal coordinator for each community, (3) accelerated reuse-driven cleanup, (4) fast-track property disposal, and (5) coordinated efforts to pool all federal resources.

ability to make a reuse-driven cleanup compatible with CERCLA, but also changes in the priorities and traditions that have long shaped the cleanup project.

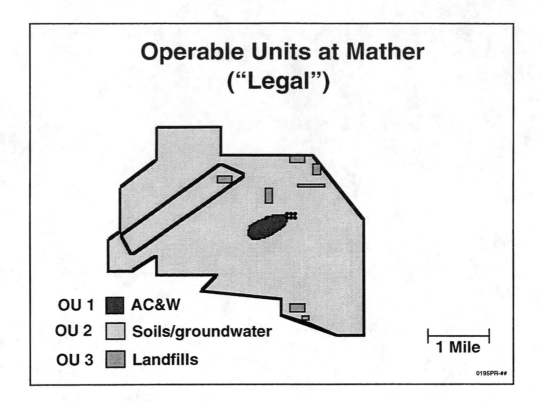

Figure 14

Figure 14 provides an overview of Mather AFB and its operable units. The runways are indicated by the nearly rectangular box. As noted earlier, there are more than 69 individual sites catalogued in DoD's IRP. The FFA lumps these sites into three operable units. OU 1 consists of one site--the AC&W site--and the associated groundwater plume. This was the first major contamination problem discovered at Mather and was the site listed on the NPL prior to the fence-line-to-fence-line listing of the entire installation.

OU 3 consists of seven landfills. OU 2 comprises 59 sites and includes all other soils and ground water problems at the base. We should note that approximately half of these 59 sites may be candidates for a "no-further-action" finding in the ROD.

Figure 14 illustrates that neither reuse nor risk was a primary consideration in the designation of the OUs. Instead, chronology, site characteristics, and convenience (from a regulatory perspective) were key factors. The AC&W site was designated as OU 1 because initially it was viewed as Mather's major contamination problem and in this sense does correspond to a risk perspective. However, designation of all the landfills as a single OU clearly occurred because of the potential economies of scale if a common remedy could be found (such as consolidation and capping at a single site). Initially, there was little evidence of substantial contamination outside OU 1 and 3; hence, all other problems were lumped into OU 2. However, significant soil and groundwater problems have now been discovered in OU 2. In fact, the groundwater plume emerging from the northern edge of OU 2 may represent Mather's most serious environmental problem.

The existing arrangement of OUs is clearly inconsistent with a reuse emphasis and only partially consistent with a risk-based approach. To summarize, OUs at Mather are based on

- the chronology for discovering contamination
- potential economies of scale
- regulatory convenience
- risk (partially).

OUs based on reuse concerns would logically take on a stronger geographical emphasis. In addition, the large size of OU 2 would seem to preclude cleanup and transfer of a limited portion of a base. Although the FFA divides the cleanup into three parts, OU 2 is so large that the cleanup involves a basewide approach, sequentially moving through the steps highlighted in the introduction.

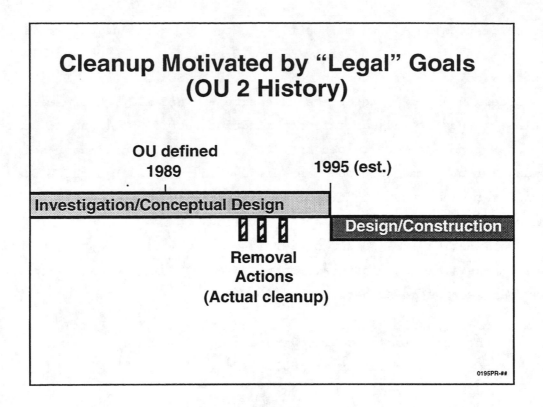

Figure 15

The basewide orientation of the cleanup efforts results in the need to investigate and assess the entire base prior to remedial action. Figure 15 illustrates the status of OU 2 and indicates that virtually no cleanup has taken place. It was only in 1994 that three small removal actions (actual cleanup efforts) were undertaken.

The project flow illustrated in Figure 15 is typical for many CERCLA-regulated projects. To avoid unacceptable "midnight" remedies, an elaborate procedure was developed to provide checks, balances, and thorough investigations before any remedy is selected. At Mather's OU 2, the result has been a series of lengthy and repetitive studies that have not yet culminated in a ROD. OU and OU 3 have moved down similar paths with a ROD completed on OU 1.

Figure 15 indicates that three removal actions were undertaken (in 1994). Removal actions, which are intended to address urgent issues, are implemented by authority of the lead agency (DoD). There are both time-critical and non-time-critical removals. The latter require an Engineering Evaluation/Cost Analysis (EE/CA). EE/CAs were submitted for the three OU 2 sites in early 1994 and, after a one-year delay, were set in motion early in 1995.

Removal actions at Mather AFB are indicative of the philosophy governing the project. The three removals were intended to generate capping material for the landfills, which are organized into a single OU to achieve a scale economy. The removal actions were not intended to address risk or reuse goals. As such they represent a continued emphasis on total base cleanup as opposed to interim goals. In addition, funding for 10 other removal actions, intended to demonstrate "speed," was denied by the Air Force Base Closure Authority in favor of ongoing OU-based monitoring.

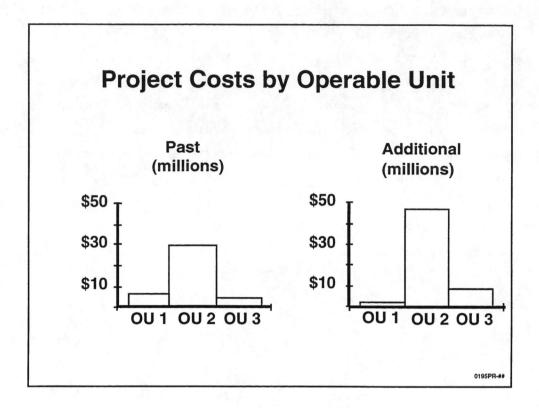

Figure 16

Figure 15 illustrates that the cleanup has followed the philosophy of the FFA, but little cleanup work has been done. Figure 16 shows the funds expended to date on the project and provides Mather's own estimates of the funds required to complete the actual cleanup work. Consistent with the FFA, estimates are divided by operable unit. It is interesting to note that funds expended to date are not significantly smaller than the projected funds needed for cleanup. In the case of Mather AFB, study and investigation costs have been comparable to estimated cleanup costs. We should, however, note that estimated costs will typically rise whereas study costs have already been expended. In addition, the ultimate achievement of groundwater standards is questionable from an engineering feasibility perspective. The above costs reflect estimates for putting remedies in place, not for ensuring that standards are achieved.

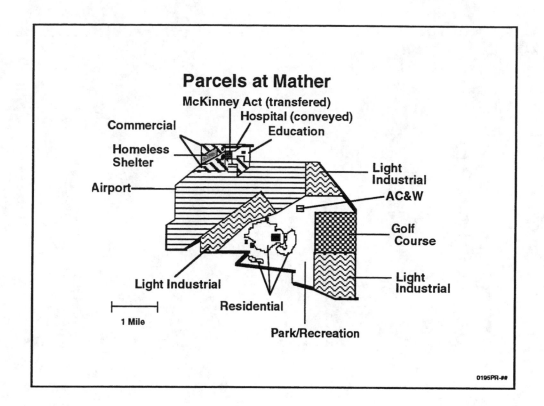

Figure 17

Figure 17 depicts parcels and proposed reuse designations at
Mather. It should be noted that reuse plans have changed over time.
For example, the light industrial reuse area depicted in the southeast
corner of the base had, at one time, been proposed for parkland or for a
second golf course. One consistent issue--since the designation of the
County of Sacramento as the reuse authority--has been the emphasis on
achieving timely reuse of the airport parcel. With the exception of
McKinney Act transfers, conveyance of the hospital to McClellan, and the
transfer of control of the Mather golf course, reuse negotiations have
focused on the airport.

Recently these negotiations have culminated in a lease agreement
with the County of Sacramento for use of the airport parcel and the two
light industrial parcels directly adjacent. The ability of the county
to attract tenants is unknown. Airborne Express is slated to begin air
cargo operations at Mather Airport in October 1995 and Trajen Flight

Services, which provides aircraft maintenance and fueling, has leased 60,000 square feet of hangar space and the fuel farm.

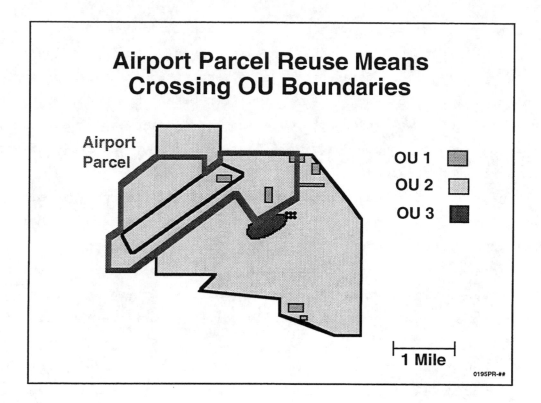

Figure 18

Figure 18 illustrates an alternative approach to the cleanup at Mather that would emphasize reuse. Rather than orient investigation and cleanup at a basewide level, DoD could take a more limited phased approach to cleanup, focusing on the most important parcels for reuse and acknowledging that other parts of the base will have to be cleaned up at a later time. These other parcels would remain in DoD hands indefinitely. This is not different from the current basewide approach to cleanup, which also requires DoD management of an indefinite length.

In Figure 18, the airport parcel is illustrative of this approach. Some reuse is occurring through lease to the county, though there are outstanding legal issues regarding the use of leasing prior to full implementation of cleanup remedies. Given these legal issues, the possibility that reuse may still require transfer by deed, and the desire to attach additional users to this parcel, one version of reuse-driven cleanup would be to focus efforts on attacking the sites within this parcel.

For illustrative purposes, we assume focusing cleanup efforts on the airport parcel now represents a reuse-driven strategy. As this is the only parcel with a foreseeable reuse, such a focus also represents one of the most important elements of a risk-based strategy. It is the only parcel where we can expect reuse and hence human exposure. Thus, on leased parcels, there is a significant overlap between reuse-driven and risk-driven strategies.

Figure 18 illustrates that the airport parcel incorporates sites from OU 2 and OU 3. The ability to focus cleanup efforts on such a parcel would require redefinition of the OUs or at least regulatory concurrence to conduct the project in a manner that is not consistent with the FFAs. Thus, while reuse-driven and risk-driven strategies may be compatible, they are not necessarily compatible with a "legal" strategy.

We should note that if the reuse prospects for a leased parcel are not affected by existing contamination, an alternative reuse-driven strategy may be to focus efforts on a parcel where transfer-by-deed is essential. Cleanup to address known exposure pathways on the leased parcel would correspond to risk-based priorities, and work on the transfer-by-deed parcel would correspond to a reuse-driven strategy. Under these conditions, the key elements of a risk-driven and reuse-driven strategy would involve work on two parcels rather than the single parcel assumed above.

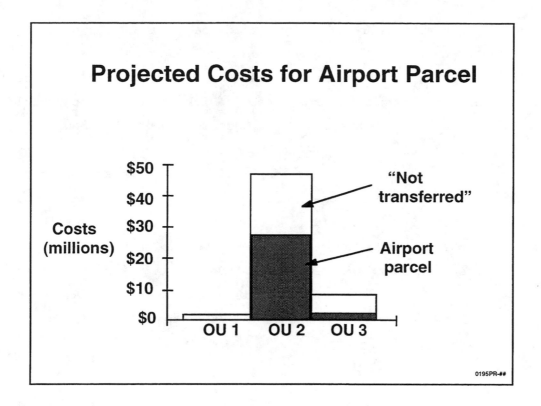

Figure 19

For purposes of analysis and to be able to apply this example elsewhere, we assume that a limited reuse-driven strategy at Mather can be accomplished by focusing on the airport parcel. This may overlap the most critical aspects of a risk-driven strategy as a result of the presence of tenants on the parcel.

Figure 19 shows that the estimated costs for cleaning up the airport parcel are only about half of the costs for the entire cleanup. As noted previously, costs for a limited reuse-driven strategy could be even lower if a smaller parcel was judged to be most important for reuse. The graph in Figure 19 is based on the assumption that the airport parcel depicted in Figure 18 represents the highest priority for reuse-based cleanup.

This assumption raises a concern regarding all other parcels. Assuming that a reuse strategy at Mather is best developed by concentrating on only the airport parcel, transfer or reuse of the other parcels would be indefinitely delayed. Estimated costs for their cleanup are illustrated by the white portions of the bars in Figure 19

and marked "not transferred." They include parts of OU 2 and OU 3 and
all of OU 1.

Under a traditional exposure assessment, these other parcels could
receive a high-risk ranking, and, hence, a risk-driven strategy could
conflict with the reuse-driven strategy assumed above. Although there
may be no planned or imminent reuse, traditional exposure assessment
does not account for the immediacy or likelihood of the assumed end-use
scenario. If these factors are included, a reuse-driven cleanup would
be nearly identical to risk-driven cleanup. This still leaves the
question of whether sites on other parcels, for which there is no
imminent reuse and which will remain in federal hands--or are leased and
monitored to avoid exposures--pending cleanup, pose risks that merit
consideration by other than an end-use scenario metric. This is
depicted in the following figure.

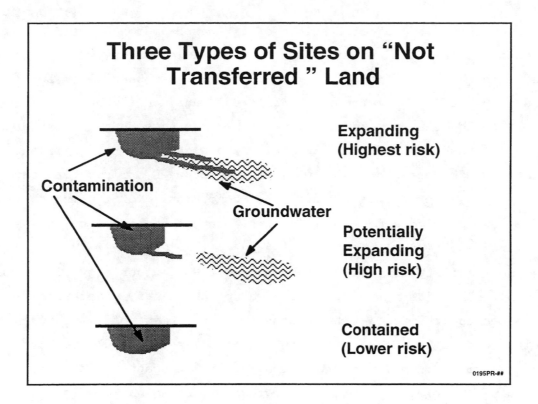

Figure 20

The limited reuse-driven cleanup strategy illustrated in Figure 18 addresses all sites within one reuse parcel, but it is not necessarily consistent with a risk-based approach for the entire base. This raises questions about how risk is determined and the meaning of a risk-based strategy. For parcels that are the focus of reuse efforts, the meaning is straightforward. There is an end-use scenario that gives meaning to a risk calculation. For sites where there are no imminent plans for transfer or reuse by lease, the end-use scenario is less meaningful even though the scenario is used in traditional risk ranking.

Figure 20 categorizes sites according to their potential for contamination to spread. We suggest that the potential for spread is a far more meaningful way to rank risks for those sites that will remain in federal hands (or are subject to federal monitoring in a lease condition), without reuse and pending cleanup. Obviously of greatest concern is the potential contamination of ground water. Should this occur, cleanup costs will escalate, and there may be few technical options for achieving State-of-California ground water standards.

With this site categorization, a risk-based approach could be harmonized with a limited reuse approach by giving priority to sites outside the key reuse parcel(s) with the greatest potential to spread. Sites with little or no potential to spread, and not in the key reuse parcel, could be addressed at a later time. We note that this strategy is possible because emergency removals have already cut off imminent threats to human health and known exposure pathways prior to lengthy CERCLA and SARA protocols. Under these conditions, DoD can monitor and control exposure on sites where it still manages the property unless the site has the potential to spread.[12]

The viability of this approach as a risk-based strategy lies with the many sites within a base where cleanup can be safely deferred. This arises because the fence-line-to-fence-line treatment of military bases results in detailed regulatory scrutiny of sites that would be given far less attention if they were part of a pattern of scattered private ownership. Many sites have small degrees of contamination--low potential for spread--but receive the same priority as more-complicated sites when they are lumped into large OUs. Typically, it is a small number of sites that leads to regulatory attention for the entire base.

An additional reason for categorizing risk by spreadability is the recognition that budgetary limitations imply that cleanup programs will last many years, if not decades. During this time, the federal government will manage the property and can take measures to ensure that the public is not exposed to contamination if the site is not spreading. During the period of federal ownership and management, spreadability is a far stronger indicator of risk than exposure at the site. When the site is eventually transferred, such an assumption may no longer be true.

We emphasize that the issues of a risk-based approach for a site still in federal hands (and eventually cleaned up) differs greatly from the issues associated with minimizing risk after property transfer. In

---

[12]Section 300.430(a)(1)(iii)(D) of the NCP states that USEPA expects that institutional controls will be used during the course of the remedial project. Obviously, the lengthy times required to complete cleanups on closing bases under conditions of limited funding were never anticipated.

the latter case, a highly toxic site, with limited spreadability, could pose a significant risk because the federal government cannot guarantee the new owner's vigilance in ensuring that individuals are not exposed to harm because of contaminated soil. However, if there are no immediate plans for property transfer, it is those sites that have the potential to create hazards in unforeseen places that represent the greatest risk.

Finally, we note that this approach to risk assessment can lead to considerably different priorities than traditional exposure assessment. Under traditional exposure assessment, a site with more-toxic materials, but less potential to spread, might be viewed as a higher-risk site than a site with opposite characteristics. As mentioned above, when reuse is at hand the traditional approach becomes more appropriate. However, when reuse is too distant to contemplate--and the sites remain under federal management--it is the possibility of institutional controls failing (spreadability) that is a better determination of risk.

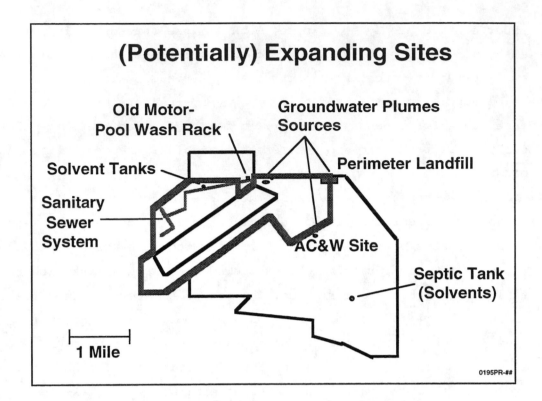

(Potentially) Expanding Sites

Old Motor-Pool Wash Rack

Groundwater Plumes Sources

Solvent Tanks

Perimeter Landfill

Sanitary Sewer System

AC&W Site

Septic Tank (Solvents)

1 Mile

0195PR-##

Figure 21

Figure 21 illustrates those sites at Mather that are expanding or have the potential to expand. Many of these sites are within the boundaries of the illustrative reuse parcel while others are not. The "risky" sites involve all three OUs.

The issue arises of how DoD should determine which sites are unlikely to spread and therefore can be remedied at a later time. One test involves the current status of these sites if they have existed for decades and have not spread. In addition, straightforward parameters like soil permeability, viscosity, and distance to groundwater provide an engineering scale for reaching judgment. Obviously, there is a continuous scale and a ranking; however, the sharp difference among many contaminants, such as solvents (which can be assumed to have a spread potential) and lubricants, makes for well-defined distinctions. Although many may still feel that errors can be made and "contained sites" may ultimately spread, current CERCLA protocols also fail to address this concern, and budget limitations imply that such choices about priority in cleanups will be made.

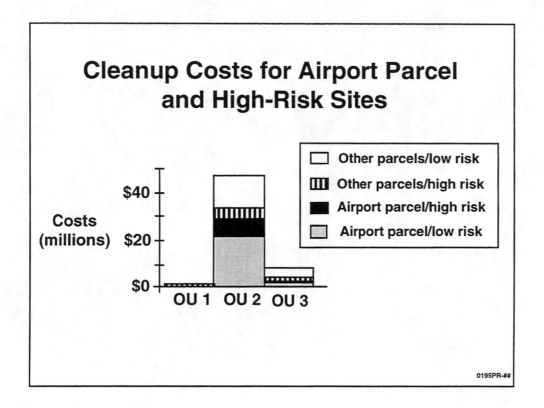

Figure 22

Figure 22 illustrates the costs of addressing sites on the airport parcel and the sites where contamination could spread (labeled high risk). The black part of the bar represents the costs of remediating sites in the reuse parcel that also have the potential for spread.

The key point of Figure 22 is that a dual goal of limited reuse and cleaning up sites with spread potential can be achieved at far less cost than total cleanup. Had we picked a smaller reuse parcel, then costs would have been even smaller. The existing OUs provide no guidance on how to focus cleanup efforts in times of limited budgets.

The costs in Figure 22 represent future costs. An interesting point is the extent of past costs (the left side of Figure 16) that could have been avoided had there been early identification of interim goals. The basewide emphasis has pushed projects toward comprehensive and detailed investigations. This was particularly prevalent at a time when it was expected that bases would be quickly transferred in their entirety. We now know that, in many cases, transfer by deed will proceed slowly, one individual parcel at a time. Even in a lease situation such as at Mather where much of the base may be leased to a single entity, reuse may still mandate a sequenced cleanup strategy. We suspect that the early identification of interim goals will allow for significant reductions in basewide investigation costs.

Mather Summary

- **Project following "Legal" OU-based path**
- **"Legal" means protocol driven**
  - not risk driven
  - not reuse driven
  - not speed driven
- **Interim goals could be partial reuse and cleaning up risky sites**
- **Internal base boundaries would have to be redrawn**
  - OU should conform to parcel
  - BCT could facilitate

0195PR-##

Figure 23

Figure 23 summarizes our findings for the cleanup project at Mather AFB and is typical for many other bases. The project is following a compliance-based path, which organizes the program around three OUs. The approach appears to be focused on achieving regulatory milestones rather than on minimizing risk, maximizing reuse, or speeding cleanup.

Our analysis suggests that a limited reuse goal--along with strategy of cleaning up other sites where there is a potential for contamination to spread--may represent a means of blending reuse-based and risk-based strategies under budgetary constraints. However, this will require adjusting internal base boundaries defined by OUs. The BCT has the authority to make such changes, but the need for experience with this process, the need to overcome administrative detail, the costs of changing project direction, and lack of clearly defined interim goals reduce the incentives to do so. DoD should clearly state that the goals of cleanup are related to policy objectives, such as reuse or risk reduction, rather than to simply satisfying administrative milestones or seeking total cleanup.

To summarize, a reuse-driven strategy, a risk-based strategy, and a "legal" strategy can be harmonized on a large base by following the steps listed in Table 2.

Table 2

Steps for Harmonizing Multiple Goals Under Limited Budgets

1. Focus remedial action on parcels where cleanup directly facilitates reuse. This may involve a total parcel focus to prepare for transfer by deed, or a focus on tasks that attract tenants to leased parcels.
2. With remaining funds, focus remedial actions on all sites across the base with significant potential to spread.
3. Defer cleanup for sites that meet two criteria: (1) there is no immediate reuse scenario and (2) institutional controls under federal management effectively reduce risk.
4. Renegotiate FFAs/FFSRAs for consistency with steps 1-3.

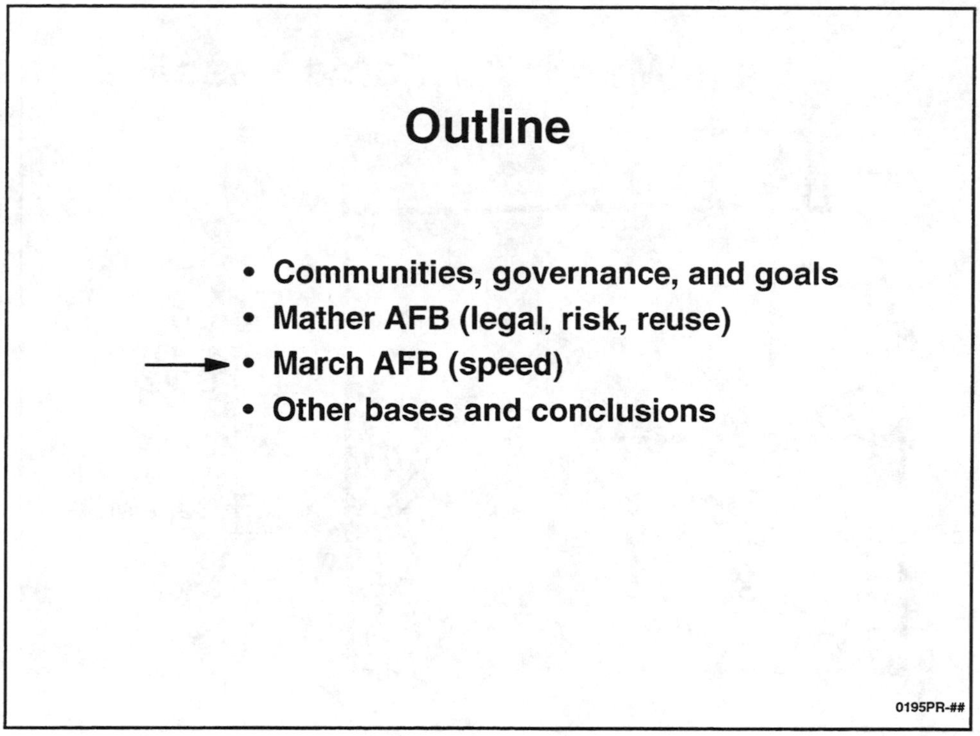

Outline

- **Communities, governance, and goals**
- **Mather AFB (legal, risk, reuse)**
→ - **March AFB (speed)**
- **Other bases and conclusions**

0195PR-##

Figure 24

In the following pages, we discuss the cleanup project at March AFB. This project indicates methods of increasing cleanup speed and overcoming barriers when OUs do not correspond to interim policy goals.

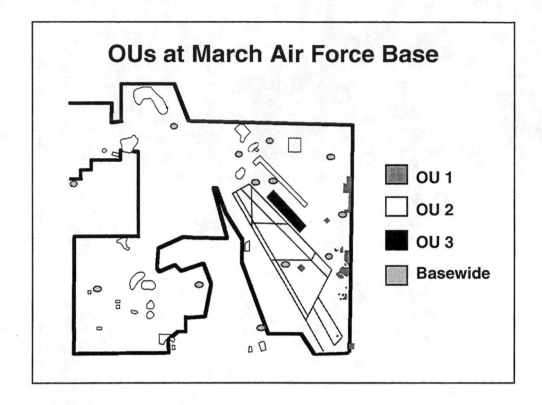

Figure 25

March AFB is located approximately 75 miles east of Los Angeles near the city of Riverside and the Moreno Valley. Throughout the 1980s, Riverside County was one of the fastest growing regions of the nation, though the recent recession has slowed economic growth. Community interest in economic reuse of the base follows a similar pattern as that for the reuse of Mather; over the long run, real estate is likely to be highly desirable, but near-term interest has been dampened. Only half of March AFB is being closed. A portion of the base will remain open for a reduced mission to be performed by the Air Force Reserve and Air National Guard.

Figure 25 highlights the OUs at March AFB. The portion of the base east of the narrow neck (the portion containing the runway) will remain active, while the western half of the base will be closed. The eastern half of the base contains most of the base infrastructure, while the western half is largely empty fields that have been contaminated over many decades. New residential neighborhoods border the base on the west. Cleanup for the entire base is funded from the BRAC account.

OU 1 and OU 3 are related to specific "hot spots" and have a reasonably sharp geographical definition. However, OU 2 was created to include all other sites at March crossing both the closing and active portion of the base. Clearly the OU was not intended to facilitate reuse. An additional OU (called basewide or 2a), which covers disparate areas of the base, has been created for sites found after OU 2 was defined. Thus, two separate OUs span various pockets of the entire base. Given the dispersion of OUs 2 and 2a (basewide), the complete process for both OUs would have to be brought to the end points before most parcels could be cleaned. Thus, a project based on the OUs would lead to a similar basewide emphasis as that for Mather.

One interesting anomaly is that part of the March AFB NPL site is a very small parcel located approximately 100 miles north of the base in the Mojave Desert. This is a by-product of fence-line-to-fence-line listing.

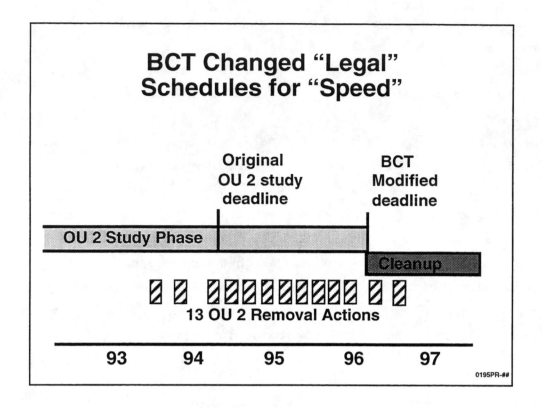

Figure 26

Although the design of OUs at March AFB present similar problems as those at Mather AFB, Figure 26 indicates that the approach to cleanup at March AFB has made this problem largely irrelevant. At March AFB, the BCT has adjusted the schedules for the legal protocols to allow cleanup to proceed through DoD's authority to implement removal actions.

Figure 26 shows that approximately 13 removal actions will have taken place at OU 2 by roughly the same time that the RI/FS will be completed. In other words, much of the cleanup may be complete by the time the CERCLA and SARA protocols determining the cleanup remedy are formalized. This was accomplished for OU 1, whereas the ROD signed in 1994 merely formalized the ongoing remedial action.

We note that the cleanup at March has not really been reuse driven since the emphasis has been to clean up as much of the base as possible. However, speed-driven emphasis may be the most appropriate term, given that only half of March is closing and there is no overarching reuse concept. Should funding for all the removal actions not be forthcoming, March may need to make choices such as those we outlined for Mather. It

is unlikely that the current definition of OUs will help in those
decisions.

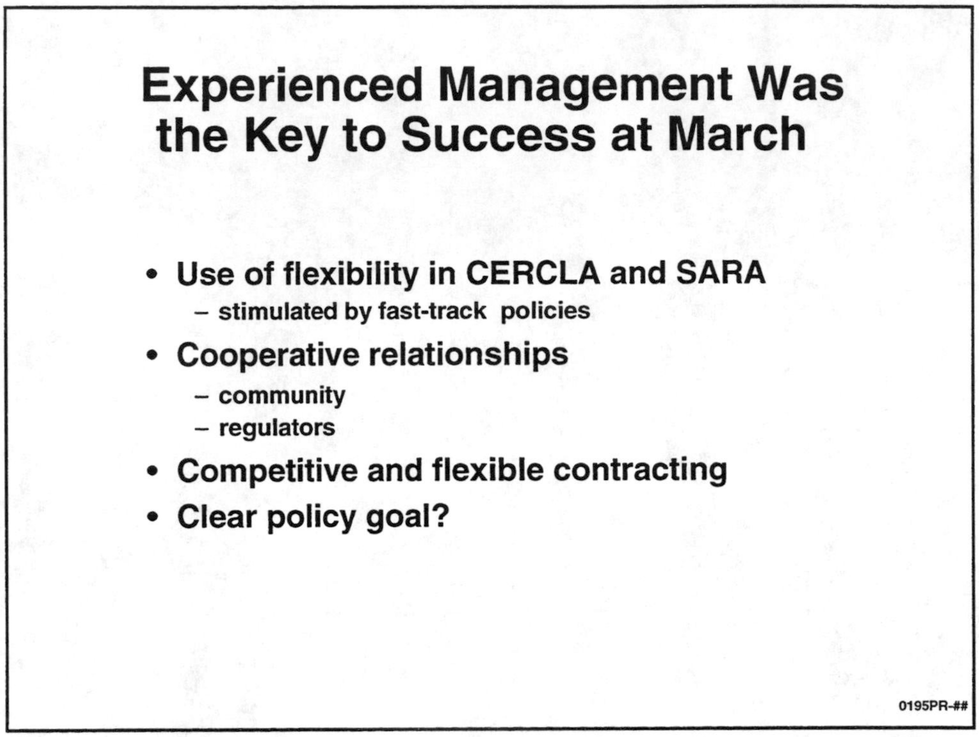

**Experienced Management Was
the Key to Success at March**

- **Use of flexibility in CERCLA and SARA**
  – stimulated by fast-track policies
- **Cooperative relationships**
  – community
  – regulators
- **Competitive and flexible contracting**
- **Clear policy goal?**

0195PR-##

Figure 27

There are numerous attributes of the cleanup at March AFB that have allowed this accelerated model to unfold. Generally, we argue that it is the skill and experience of project managers that was responsible for the successful implementation of a speed-driven approach. Key attributes of this management are described in Figure 27.

Much of March Air Force Base will be cleaned prior to completion of the CERCLA protocols. Obviously, project managers were able to take advantage of the existing flexibility in CERCLA and SARA (Table 1). Most obvious was the aggressive and creative use of removal actions, which will be discussed in the following chart. In addition, project managers took advantage of virtually every item listed in Table 1 to move the project toward completion. One interesting example relates to the level of detail required for studies specified in CERCLA and SARA (item 3 in Table 1). For the removal action on site 40, March AFB submitted a dramatically condensed Engineering Evaluation/Cost Analysis (EE/CA), which is required to comply with the NCP for non-time-critical removals. The documentation was adequate for BCT regulators as well as for other state and federal officials who coordinate permits. The project obtained a 404 permit (wetlands) and other state permits within 30 days.

Significant credit should be given to the fast-track policy for creating a climate of innovation and flexibility. The BCT concept created a mutual understanding of goals and values among DoD and the regulators that was consistently translated into rapid regulatory review and flexibility. We should also note that the Riverside/Moreno Valley community is generally conservative and has not voiced significant distrust of the Air Force. However, a similar conservative community surrounding nearby Norton AFB has been far more vocal.

In addition to flexibility, the project team had far more experience than many teams at other bases. This was critical because the numerous removal actions could be carried out only by weaving through the DoD funding and contracting system in creative and innovative ways. March personnel forced DoD's contracting service centers (i.e., the Air Force Center for Environmental Excellence and Districts of the Army Corps of Engineers) to compete with each other in terms of costs and schedules and thereby accelerated the entire process.[13] March has even stimulated the competition further by utilizing the Idaho National Engineering Lab and contemplating the use of the USEPA for contracting services. These competing service centers have brought a diverse set of contractors to March, all of whom are coordinated by active DoD project management.

It is interesting to note that this contracting model stands in direct contrast with what is becoming DoD's preferred approach of relying on large regional contractors who conduct entire cleanups at several bases in a region and are given significant responsibility for implementation of the overall project. The March AFB models suggests that administrative economies of scale associated with regional

---

[13]March personnel cite an example of a data collection task required for a removal action on OU 3. Competing service centers offered work schedules ranging from 32 days to 6 months and times to start work ranging from 14 to 87 days.

contractors may be less important than creating a competitive environment where the DoD project manager actively engages as the "general contractor."

One of the interesting questions is whether or not a clear policy goal is necessary to motivate this type of success. At March, reuse is not an overriding concern because only half the base is closing. March personnel have responded to the request by policymakers to speed cleanup rather than to achieve specific interim goals. With limited budgets, the challenge will be to transfer this model to other bases and to apply it in a more focused manner.

March AFB Uses Removal Actions Creatively

- **DoD has authority for "urgent" action**
  - "time critical"
  - not "time critical"
- **BCT may accept removals as final actions**
- **Citizens have not challenged, but . . .**
- **BCT/RAB adjusted "legal" schedules**
  - OU may be clean before "legal" studies are done

0195PR-##

Figure 28

Figure 28 highlights use of the removal action policy that led to successful efforts at March.

The BCT has given less priority to the required protocols than to the tasks required to clean up the base. They have done this through creative use of DoD's removal authority in CERCLA. These actions can be implemented unilaterally by the federal agency that is the responsible party and are intended for "urgent" action. Nonetheless, the NCP still allows for non-time-critical removals, which can require as much as a six-month planning phase.[14]

The ability to use removal actions as a basis for overall cleanup creates opportunities and problems. There are opportunities because DoD can undertake cleanup without deferring to CERCLA and SARA protocols. But difficulties arise because removals are not legally mandated and require DoD to implement proactive policies. DoD's entire environmental program is well understood to be compliance driven. The approach may

---

[14]NCP, Section 300.415.

also stretch the bounds of the originally intended meaning of the removal policy though it does appear to be consistent with changing regulatory guidance.[15] Unless regulators are willing to certify the action as a final remedy, the site could still be viewed as an unremediated site. Regulators maintain the authority to certify the removal as the final required action in the ROD. Communities can also insist that FFA milestones be implemented and could gain enforcement authority through the courts.

At March AFB, state, local, and federal regulators and the community have accepted the use of removal actions as the core cleanup strategy. At times, regulators have even threatened to return the project to the traditional "protocol-driven" approach if Air Force funding for the accelerated program was not forthcoming.[16]

Clearly, some aspects of the March model can be translated to other bases. However, the use of removal actions has been an issue that requires greater clarification, at least at the policy, political, and program management levels. The USEPA as early as the late 1980s in its Superfund Accelerated Cleanup Model (SACM) began to hint at use of removals as a means to speed cleanup as does the previously referenced guidance. Nonetheless, removals are not widely understood to be an option for the core of a cleanup strategy. Clearer policy guidance, or perhaps a change in the NCP, may be required to expand use of the March model. At a minimum, project leaders with experience and the desire to

---

[15]See for example, "EPA Guidance on Evaluating the Technical Impracticability of Ground-Water Restoration," OSWER Directive No. 9234.2-25, Section 2.2.1. Also, see the August 22, 1994, memorandum from Steven Herman, USEPA Assistant Administrator, Office of Enforcement; Elliott P. Laws, Assistant Administrator, Office of Solid Waste and Emergency Response; and senior DoD and DoE officials to the Waste Management Division Directors of the 10 USEPA regions as well as to other senior DoD and DoE personnel. This memorandum seems to encourage the use of removal actions in much the way March AFB has used them.

[16]This controversy involved a removal action at site 6 and was largely motivated by a delay in reviewing a plan to correct improperly analyzed laboratory data. The Air Force Center for Environmental Excellence (AFCEE) at Brooks AFB took six months to approve the plan. The regulators' insistence on funding was driven by a desire to overcome this delay.

utilize all available tools are required. In addition, interim policy goals that require cleanup (as opposed to studies) may be needed.

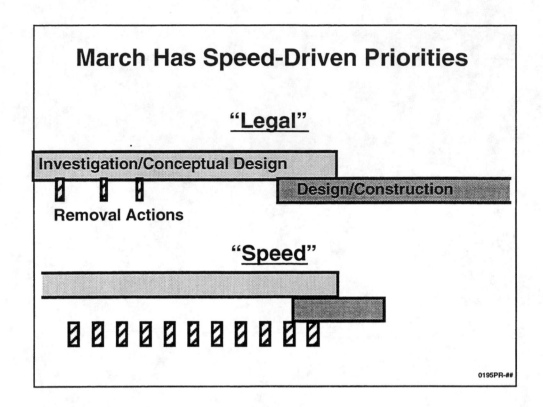

Figure 29

Figure 29 summarizes the approach taken at March. Rather than conducting a protocol-driven cleanup, March has conducted a speed-driven cleanup. Goals are not focused on selected parcels, but on cleaning as much of the base as rapidly and efficiently as possible. Reuse is not a strong driver, possibly because half of the base will remain open. Nonetheless, the approach used at March could be used to rapidly undertake the operations needed to make individual contaminated parcels suitable for transfer.

We should, however, note that the continued success of March's program could be jeopardized by the transfer of the base from the Air Mobility Command to the Base Closure Authority (BCA). March's RPM retired recently, partly in response to this transfer and the unwillingness of BCA to protect the ongoing approach. The key issues are (1) that transfer of the major command may also imply shifting of personnel and (2) BCA may insist on the use of a single large regional contractor. The idea that transfer to BCA would imply an entire shift in the cleanup project team was probably not foreseen at the time of

BCA's creation. It seems imperative for the Air Force to correct this policy and promote coherent staffs at bases being transferred to BCA. It also seems unwise to insist upon a large regional contractor in a situation in which competition seems to have a positive effect.

Outline

- **Communities, governance, and goals**
- **Mather AFB ("legal," risk, reuse)**
- **March AFB (speed)**
→ **Other bases and conclusions**

0195PR-##

Figure 30

Next, we examine cleanup projects at other bases and summarize our findings.

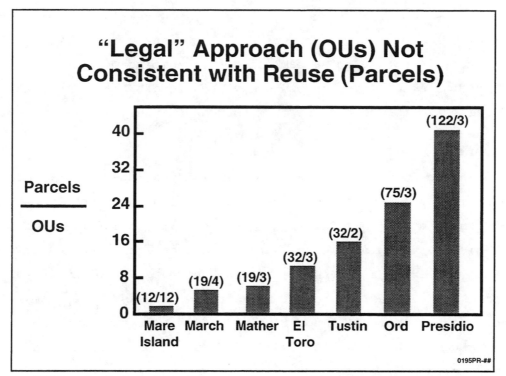

"Legal" Approach (OUs) Not Consistent with Reuse (Parcels)

NOTE: Fort Ord actually contains two OUs and numerous individual sites. The FFA called for efforts to be organized around OU 1, OU 2, and all other sites at the base, thus in effect creating a third very large OU. A 10-volume draft basewide RI/FS was issued in August 1994, *Basewide Remedial Investigation/ Feasibility Study, Fort Ord, California, Draft*, U.S. Army Corps of Engineers.

Figure 31

The discussion of the cleanup project at Mather AFB illustrates broad trends present at other California closing bases. March AFB is an outlier that may contain lessons for innovation.

Figure 31 illustrates the ratio of parcels to OUs at March, Mather, and five other bases. The chart illustrates that each base has a high ratio of parcels to OUs. The exception is Mare Island Naval Shipyard, which was in the unusual situation of being a BRAC III base with a new cleanup program and having an approved reuse plan. However, most bases have a small number of OUs and are moving down a basewide approach, with parcels having a minimal effect on cleanup. We caution that the ratio of parcels to OUs is not the only relevant factor. A base with a single OU and a single parcel would still involve a basewide approach to cleanup. What is required is a definition of OUs that is consistent with the reuse concerns at the base. We do, however, note that larger numbers of OUs imply larger administrative burdens under CERCLA; hence, there is no alternative to streamlining this cumbersome process.

Each of the following bases involve challenges that amplify the issues raised in the discussion of Mather and March.

MCAS TUSTIN

The Marine Corps Air Station is a small site (1595 acres) located approximately 40 miles south of Los Angeles. Tustin is not listed on the NPL, and the state of California is the lead regulator. As a smaller site, a basewide cleanup strategy poses less of an obstacle to reuse than at larger sites. Nonetheless the reuse authority (the City of Tustin) has divided the base into 32 reuse parcels. The base is also divided into two OUs. OU 1 consists of groundwater contamination underlying the base. OU 2 consists of all soil contamination and is scattered across the base.

Tustin is a BRAC III base, and the later listing allowed for a more explicit recognition of reuse in the cleanup strategy. Despite these apparent advantages, the process at MCAS Tustin to date has not been particularly successful. For example, four parcels at the base were declared "CERFA clean" and suitable for transfer by the Marine Corps. However, state regulators insisted that these parcels be "proven" clean. Since these parcels were not of immediate concern for the reuse authority, and because of the ongoing characterization of a plume of contamination in groundwater underlying the base, it is unlikely that these parcels will be available for reuse in the near term.

Marine Corps and Navy personnel also have attempted to initiate removal actions at MCAS Tustin in an effort to accelerate the cleanup process. However, these efforts have been delayed by regulators who have focused on the characterization of groundwater contamination at the base as their first priority. One removal action did take place at the base in 1987, and two other emergency removals have been undertaken. However, none of these actions are expected to speed up the remediation process of the overall base.

Finally, dividing the base into small components may potentially bypass certain financial economies of scale. Tustin has purchased a $7 million mobile soil treatment unit to be used for basewide (i.e., OU 2) soil remediation. If cleanup were to be focused on a small portion of the base (e.g., a high-priority reuse parcel), this economy of scale might not be achieved. Efforts to achieve interim goals at larger bases will need to be scrutinized for lost economies of scale. However, given current budget trends, it may be difficult to achieve many economies of scale.

FORT ORD

Unlike both Mather and March, Fort Ord has been and continues to be the object of intense public, political, and regulatory scrutiny that acts to slow cleanup. Located just to the northeast of Monterey, the site is widely considered to be one of the most beautiful spots in California.

The cleanup of the more than 20,000-acre base is divided into two small OUs and more than 40 individual sites that are not grouped into OUs. Although this division should facilitate aligning internal boundaries with reuse considerations, to date the cleanup has been conducted at a basewide level. A draft RI/FS for the entire base was delivered in August 1994, four years after the FFA was signed and one year after the original deadline. The document is still under regulatory review.

Primarily due to the political initiative of then-Congressman Leon Panetta (currently President Clinton's White House Chief of Staff), several relatively small parcels of the base were transferred on 29 August 1994 to the California State University system and to the University of California system. Although legal transfer of the property was accomplished on the date previously indicated, ceremonies celebrating the transfer took place on 8 July 1994.

Extensive delay in transferring the property was caused by regulatory pressure to satisfy the legal protocols for cleanup and transfer. In particular, Fort Ord BRAC personnel were required to certify that the parcels to be transferred were clear of unexploded ordnance. Reportedly, USEPA and Cal-EPA regulators were not provided with adequate time or materials to review and assess the environmental condition of the property.[17] But, under the political pressure of a planned presidential visit (later canceled) and a visit by the Secretary of Defense, agreement for transfer was reached. However, it is doubtful that this could serve as a model for other bases.

The cleanup at Fort Ord also illustrates that a speed-based strategy, similar to the one used at March, may be difficult to implement. A total of 14 sites at Fort Ord have been proposed for "interim action." Specifically, sites documented to have limited soil contamination have been categorized for early action. However, unlike March AFB, where removal authority has been used for such sites, the Fort Ord BCT has pursued remedial action by designating these sites for interim action. However, it has taken almost three years from proposal to the initiation of cleanup.

Efficiency through the use of interim actions purportedly is achieved through a so-called "plug-in" ROD, which is prepared for sites with common characteristics--limited soil contamination in this case. More-specific information is then inserted into this template document for regulatory and public review. In addition, a common or centralized remedial approach may be employed for all of the sites. In practice, however, efficiency in terms of speed of review and implementation has not been realized. The interim actions were first proposed in early 1993.[18] The first interim action remedial activities began in June 1995.

Cleanup and reuse at Fort Ord are also hampered by the issue of unexploded ordnance (UXO) and ordnance wastes. A large part of the base is commonly acknowledged to be at risk for such contamination. Further, certain parties allege that the UXO problem is much more widespread throughout the base. Regulatory decisions regarding the status of UXO

---

[17]Communication from USEPA project manager.

[18]Mention of the proposed interim actions was documented in the minutes of the Fort Ord Superfund Project Technical Review Committee Quarterly Meeting, dated 28 July 1993.

and ordnance wastes at Fort Ord, as well as public reaction to such decisions, will have national implications.

THE FORMER HAMILTON AIR FORCE BASE

Efforts to clean up and transfer this roughly 1,500-acre site located about 20 miles north of San Francisco on the San Pablo Bay highlight both the highs and lows of the base closure process. The most striking aspect of the Hamilton closure is the division of the cleanup effort into three separate projects, one managed by the Navy and two managed separately by the Army. The Air Force has also provided funding for the cleanup of a landfill.

The fragmentation dates back to 1974 when the Air Force began to excess the property. There was a transfer of 380 acres to the General Services Administration (the GSA parcel) for sale and about 400 acres to the Navy for housing; the Army accepted about 700 acres and renamed the facility the Hamilton Army Airfield. Since 1975, there has been a steady decline in military activities, with most operations having ceased by 1993. The Army and Navy parcels are now being closed under BRAC.

Fragmentation and a declining military role have made the facility a symbol of neglect often used to highlight the futility of closing and transferring military bases. The GSA parcel has been excessed for more than 20 years and, despite the vibrant Marin County economy, has remained unused. Previous Army efforts to declare the parcel clean and suitable for transfer have been rejected by both the community and regulators.

Despite this history, recent events have dramatically altered the picture. Under intense political scrutiny from California senators Boxer and Feinstein (Hamilton was part of Senator Boxer's district when she was a Congresswoman) and pressure from New Hamilton Partners, the private developer hoping (and willing) to acquire part of the property, the Army began focused efforts in 1994 to clean and transfer a portion of the GSA parcel. On 6 July 1995 ceremonies were held to formalize the transfer and certification of GSA-Parcel-Phase-1A as clean by the state, which is the lead regulatory agency. The roughly 100 acres represent one of the few examples in California in which a contaminated parcel on a military base has been cleaned and transferred.[19]

The success seems to be rooted in the political pressure that was translated into an organizational commitment on the part of the Army. However, regulatory flexibility and teaming also played a role. The process was not hindered by OUs or a preexisting cleanup strategy that was inconsistent with reuse. There had never been an FFSRA, and according to one regulator, there was early recognition that such an agreement might place constraints on a project for which priorities had not been firmly established. Parcel boundaries were adjusted at several points in the cleanup effort, and there was no corresponding need to alter a complicated agreement.

As at March AFB, the remediation of what is now known as GSA-Parcel-Phase-1A at Hamilton was accomplished though removal actions. Similarly the CERCLA protocols, to the extent they will actually be fulfilled, will largely be a formality confirming the completion of the project through removal actions.

To our knowledge, Hamilton is one of the few examples of a parcel-focused cleanup in the state. As in the preparation for transfer at Fort Ord, an unusual level of political attention was required to reorient the project toward this goal. The policy challenge is to achieve this focus on a routine basis.

An additional question for DoD involves the desirability of continuing the current fragmented approach at Hamilton. For the City of Novato, Marin County, and the State of California, the former Hamilton Air Force Base is one entity, though they have all adapted to DoD's triad of projects. One individual involved in reuse planning suggested that resources might be better spent by focusing cleanup efforts on reuse GSA-Parcel-Phase-1b rather than by dispersing resources across three separate projects. The financial and schedule costs of transitioning to a unified project should be explicitly compared with the benefits of increased focus.

----

[19]The GSA parcel was created before the BRAC law, and CERFA is not applicable.

DoD should also evaluate the process currently being used to implement a removal action on Landfill 26. A special arrangement written into the Defense Reauthorization Act allows future site developers to assume responsibility for the implementation of remedial action. There are conflicting views about the desirability and effectiveness of this initiative. An objective evaluation should help DoD determine if this model should be utilized in other locations.

EL TORO

The cleanup of the El Toro Marine Corps Air Station has become somewhat notorious for the lack of progress as well as for the intense community debate about how the site is to be reused. Because the base is located in a part of Orange County where property values are high, debate about whether it should become an airport continues and may not be resolved for years. Orange County's recent financial debacle and resulting bankruptcy declaration has further delayed consideration of reuse. As of July 1995, the reuse authority had been dissolved.

The cleanup at El Toro nevertheless contains lessons regarding the division of cleanup responsibilities at a base.

The project is divided into three OUs. OU 1 includes the plume of contaminated groundwater; OU 2 includes potential soil source areas for that groundwater contamination; and OU 3 includes all remaining IRP sites. All three OUs span a significant portion of the base rather than providing a geographical focus.

El Toro's basewide OUs differ from basewide OUs at other installations because they are largely a product of contracting rules. Typically, basewide OUs have been defined by contamination type or for potential economies of scale. El Toro's OUs resulted from the financial limits of a contract. The Navy was forced to issue a separate contract to investigate and remediate groundwater contamination when the existing contract for investigation of soil source areas could no longer be supplemented. Definition of the OUs along these contractual lines followed. As a result, the division makes little sense in either a reuse or technical sense.

Eight removal actions have been proposed and implemented. However, these actions, unlike the majority of those carried out at March AFB, were used for true urgent situations and not as a basis for implementing the overall project. Specifically, they were used to cut off sources of ongoing leaks or to prevent the imminent spread of contaminants located near the ground surface. In other words, these were carried out as emergency removals that were designed to remove an imminent threat to human health and/or the environment and were not designed to speed up the remediation process.

HUNTER'S POINT

Cleanup of the abandoned shipyard at the southern edge of San Francisco has illustrated that the current "protocol-driven" approach is unlikely to create local jobs through the cleanup itself. The community of Bayview surrounds the shipyard, and there has been intense community pressure on the Navy to ensure that some of the cleanup funds result in jobs for Bayview residents. Despite some good-faith efforts by the Navy, there has been minimal success.

There are numerous reasons for the lack of success; however, one reason is that there are few jobs at the site resulting from the cleanup. Forthcoming RAND research will discuss Hunter's Point more thoroughly and illustrate that although more than $50 million has been spent on the Hunter's Point cleanup, only $2 million has been spent at the site.[20] The "protocol-driven" approach lends itself to "white-collar" studies that can be conducted at the contractor's office and lab work that can be conducted remotely. If DoD seeks to use the cleanup itself as a job-creating engine for communities affected by closure, it will have to change the cleanup strategy so that the project generates jobs at the site. In addition to reuse-driven or speed-driven approaches, another option may be to have a jobs-driven approach, emphasizing cleanup of those sites with significant manpower requirements.

---

[20]The data were supplied by the Navy and illustrate that 26 percent of the $50 million has been spent on fieldwork. However, within this category, Navy data indicate that most funds are used for administration and off-site waste disposal.

THE PRESIDIO OF SAN FRANCISCO

No closure cleanup has received more public and political scrutiny than the Presidio. Located in the Northwest section of San Francisco, and containing both a western and northern shore, the Presidio may be one of the most valuable real estate parcels in the world. It is also located in one of the most politically active communities in the nation. The Presidio is not an actual closure but a transfer to the Department of Interior with the Army retaining responsibility for the cleanup. As such, reuse can occur without the need for the complete cleanup needed for transfer by deed.

Investigations began in the late 1980s and are still under way. The base is divided into three OUs: the Public Health Service Hospital, a small area near the Golden Gate Bridge, and the main post, which constitutes the bulk of the base. Only four removal actions have been undertaken. Approximately $50 million have been expended on cleanup of the Presidio.

The Presidio is not on the NPL list, which makes the State of California the lead regulator. Typically this would imply governance by an FFSRA; however, the Army has not agreed to schedules and milestones. Currently the state expects the RI/FS to be completed in the summer of 1996.

As with other bases, the cleanup has, to date, taken a basewide focus with an emphasis on studies and analysis. Oddly enough, the Army's reluctance in agreeing to an FFSRA may result in some flexibility if there is a decision to focus cleanup efforts. At the time an FFSRA would normally have been negotiated, it may have been reasonable to expect adequate funding for a basewide cleanup approach.

One option for focusing the cleanup efforts is to concentrate activities on the roughly 1,000-ft-wide strip of land that runs several miles parallel to the northern shore and north of U.S. 101. Here there is already extensive public use, and the area is widely recognized as desirable for more expanded use. Those responsible for assembling the Department of the Interior's general management plan have continually viewed the area as a high priority. However, the use of aggressive removal actions, as was done at March AFB, would come under intense public and governmental scrutiny. Although RABs, BCTs, and reuse authorities have simplified the complex climate in many situations, the intense interest in the Presidio is probably overwhelming. Instead, it may be desirable to negotiate an FFSRA that allows focus on this parcel or on whatever may represent the cleanup tasks of highest priority.

MARE ISLAND NAVAL SHIPYARD

Lying near the intersection of the San Francisco Bay and the San Joaquin Delta, Mare Island is likely to entail one of the most complex and prolonged cleanups of all California's closing bases. In operation since 1854, this shipyard has hosted a wide variety of users who have generated an equally wide variety of wastes, including solid waste, hazardous waste, munitions, and nuclear materials. Total cleanup costs to CERCLA and SARA standards are estimated at over $200 million, while other environmental cleanup efforts are estimated to cost a similar amount.

Mare Island is a BRAC III base, and the program for cleanup is in its early stages. It is only in the last 18 months that there has been a broadly based investigation of the base. The recent initiation of the cleanup efforts has served to give Mare Island a better correspondence between reuse parcels and division of the base by OUs. The project was formulated at a time when there was a realization that complete cleanup might be infeasible and that division of the base was a critical exercise. In addition, the City of Vallejo is the sole participant in reuse planning, and although the cleanup of Mare Island is extraordinarily complicated, Vallejo has been able to develop a well-defined and approved reuse plan. Ultimately, OUs were drawn by balancing the divisions in this plan with the physical considerations of cleanup. There are, however, two OUs comprising the original 24 IRP sites that do not have a sharp geographic focus.

Finally, we should mention that Mare Island workers, with support from Navy headquarters, have made special efforts to obtain environmental training and gain a share of cleanup work for themselves. The many highly trained engineers and scientists at Mare Island are well suited for the engineering studies that characterize the CERCLA and SARA process. However, more significantly, Mare Island employees now represent a base-focused competitor to the large regional contractor responsible for the overall cleanup. DoD should monitor the effectiveness of this model both as a means of retaining displaced workers and as an alternative contracting model.

Focus of Cleanup Projects

	Protocols	Reuse	Speed
Mather	Strong	Weak	Weak
Ord	Strong	Moderate	Weak
El Toro	Strong	Weak	Weak
Tustin	Moderate	Weak	Moderate
March	Weak	Moderate	Strong
Hamilton	Weak	Moderate	Moderate

0195PR-##

Figure 32

Figure 32 summarizes the discussion following Figure 31 by providing our assessment of the emphasis of several of the cleanup projects. The adjectives are not intended to represent a value judgment, but rather to provide our assessment of where the emphasis of a project lies. As noted, none of the cleanup projects at the five bases have been strongly influenced by reuse concerns. It should be noted that Mare Island Naval Shipyard has, to date, followed a basewide approach, though this project is only in the early stages of site characterization, and that federal-to-federal transfer at the Presidio makes it difficult to compare this base with those listed above.

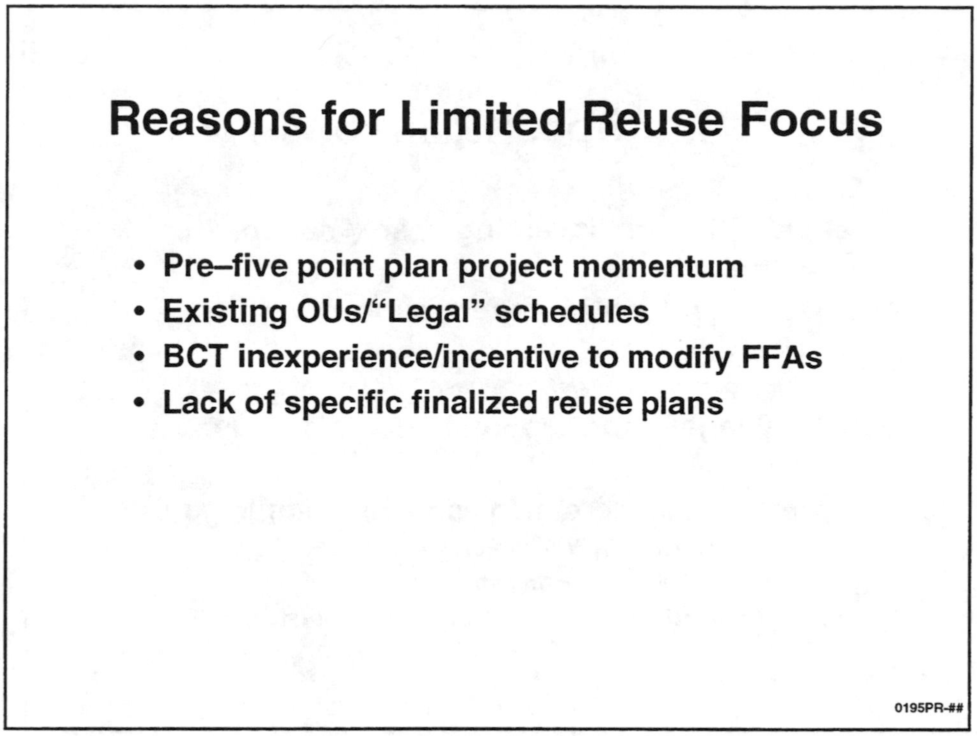

Reasons for Limited Reuse Focus

- **Pre–five point plan project momentum**
- **Existing OUs/"Legal" schedules**
- **BCT inexperience/incentive to modify FFAs**
- **Lack of specific finalized reuse plans**

0195PR-##

Figure 33

Figure 33 summarizes reasons for the lack of a reuse focus indicated in Figure 32. We note that the overall economic climate in California and the stigma associated with reusing a contaminated military base have dampened interest in reuse, independent of a cleanup strategy.

As illustrated at Mather AFB, one reason for the lack of a reuse focus was the direction of the cleanup project prior to reuse goals. This is reinforced by the regulatory obligations that remain the dominant governing mechanism. To implement a reuse-driven cleanup, the BCT must modify standard protocols. Problems in retaining experienced cleanup managers and regulators severely hamper the effort. Finally, we note that the lack of precisely defined reuse plans also acts as a disincentive to refocus cleanup projects. Communities need to provide guidance on cleanup strategy even when reuse plans are not in final forms.

Site-Specific Issues

- **What priorities are governing base cleanups?**
  - *mostly "legal" protocols*
- **Are "legal" protocols risk-reduction driven?**
  - *not necessarily*
  - *after emergency removals, "risk based" is poorly defined*
- **Do OUs ("legal") correspond to parcels (reuse)**
  - *generally not*
- **Can project managers help achieve multiple goals?**
  - *utilize flexibility in CERCLA and SARA*
  - *create competitive contracting structure*
  - *analyze to identify interim goals/new internal boundaries*

0195PR-##

Figure 34

Figure 34 summarizes the answers to the questions posed in Figure 9. "Legal" protocols, defined in the NCP, and translated by FFAs or FFSRAs, still define strategy for most cleanups. These protocols push the projects toward a basewide focus rather than toward interim goals. The protocols are not necessarily consistent with a worst-first (or risk-based) priority system. Also, emergency removal actions have eliminated known exposure pathways, and there is no obvious or widely accepted definition of a risk-based strategy.

The lack of correspondence between OUs and reuse parcels indicates that the BCT should be prepared to redraw internal base boundaries to correspond to reuse or other interim policy goals. This may need to be done at several points in a project. The ability to redraw these boundaries might require new FFAs or FFSRAs; however, renegotiation is within the authority of the BCT. The case of March AFB illustrates that with community support, the administrative aspects of redrawing internal base boundaries can be minimized.

In general, we can point to the critical role of the BCT and management of the project, and not the problems in CERCLA and SARA, as the critical factor in determining success or failure. Successful project management involves appropriate use of flexibility in CERCLA and SARA (Table 1), especially to facilitate removal actions. It also involves creation and active management of the appropriate set of contractors for a particular cleanup. This stands in contrast to current policy that assumes that large regional contractors represent the appropriate solution. This, in turn, will require active BCT decisionmaking that is typically delegated to the large contractor.

The above actions should help move a project away from studies and toward remedial action. However, even with the acceleration associated with removal actions, the problems of limited budgets and tailoring cleanup to interim goals will remain. One uncertainty in the speed-driven approach at March is that much of the base could be cleaned but few parcels will be completely clean. Thus, in addition to acceleration, analysis of the relationship among the cleanup project and multiple interim goals must take place. The elements of this analysis are described on the following pages.

Recommendations for
Multiple Goals

- **Move away from fence-line-to-fence-line listing**
  - much of CERFA is to undo the effects
- **Encourage more strategic reuse plans**
- **Divide each base by policy goals:**
  - risky sites
  - reuse-critical sites
  - other policy goals
- **Move away from current methods of dividing bases**
  - chronology, contaminant, convenience
  - achieving economies of scale

0195PR-##

Figure 35

Figure 35 summarizes initiatives and project management actions that would facilitate the achievement of multiple goals as outlined in Figure 3. As shown in the previous chart, there are significant procedural and institutional obstacles toward redrawing the internal boundaries at military bases. Any policy, such as the USEPA's fence-line-to-fence-line listing, that draws boundaries in a permanent way contributes to these obstacles. Although fence-line-to-fence-line boundaries do not preclude redrawing internal boundaries, they are psychological barriers, which can result in a significant administrative burden. Moreover, these boundaries provide false signals by placing many low-risk sites on the list of the nation's most contaminated. Much of CERFA is little more than an attempt to undo the effects of this policy.

The lack of finalized reuse plans also dissuades project leaders from reorienting cleanup toward reuse. Since reuse is affected by numerous known and unknown local zoning and planning constraints and subject to oscillations in the economic climate, it may be nearly impossible for communities to arrive at final plans in time to support cleanup goals. Instead, more strategic reuse guidance for the BCT is needed. For example, a community might inform the BCT that an area of the base will be used for commercial business prior to ensuring that all prospective businesses are identified. If cleanup is to be linked to reuse it must be done in a manner that reflects the substantial uncertainty around reuse. Cleanup projects will need to proceed on very general reuse guidance.

Most critically, interim goals for cleanups on large bases will need to be identified. Rather than moving through the CERCLA process for an entire base, project managers should conduct an exercise of dividing a base into different elements where actual cleanup can lead to a substantive achievement. Each base should be divided into those sites that are most critical for reuse, or most risky; those where cleanup would achieve other interim policy goals (i.e., job creation); and those where cleanup can be delayed. We note that for sites that have little prospect for near-term transfer, it is "spreadability" rather than toxicity that may be the most important risk factor. OUs and FFAs should be renegotiated around these goals and risk factors.

This approach stands in contrast to today's approach of dividing a base by contaminant type, chronology, or administrative convenience. Because such approaches do at times lead to economies of scale, DoD needs to ensure that division of the base has not eliminated relevant opportunities to reduce basewide costs. However, the fiscal requirements for basewide cleanup are so large that such economies, while existing, may have no policy relevance.

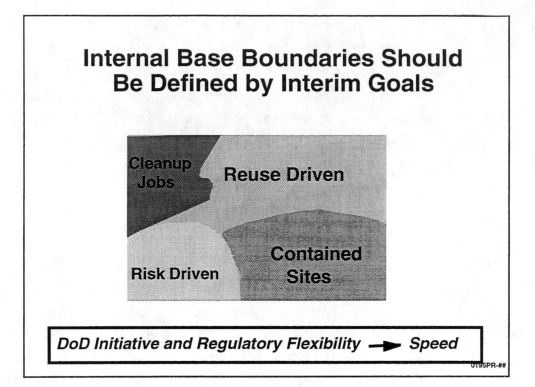

Figure 36

Figure 36 provides a graphic illustration of the recommendations discussed in the previous figure. As mentioned, the cleanup of a large base should be divided into sites corresponding to different interim policy goals. This would normally involve sites that present important risk to human health or the ecology, sites that are important for the highest-priority reuse, sites that are relevant for other policy goals, and sites that can be addressed at a later time (i.e., contained sites).

We believe there is no inherent conflict between attempting to achieve cleanups that give priority to both risk reduction and reuse. The cleanup of many military bases is so large and diverse that there are inevitably large parts of the project that can be delayed without missing important reuse opportunities or presenting undue risk to human health or the environment. The current basewide approach to cleanup does not promote finding those sites. The policy challenge is to divide such bases in ways that correspond to those considerations and to reorient cleanup projects along those divisions.

3. CONCLUSIONS

We recap by answering the four questions posed in the introduction:

1. What goals actually govern cleanup projects for California's closing bases? To what extent have cleanup projects been modified to conform to reuse-driven goals?
   CERCLA compliance, not risk reduction, reuse, or speed, is still the goal of many cleanups. A reuse goal is hampered by a poor correspondence between OUs and reuse parcels. Under an intense political spotlight, cleanup projects at Fort Ord and Hamilton Army Airfield have at least partially shifted direction toward cleanup and transfer of reuse parcels. However, this level of political attention cannot be the basis for an overall program.

2. Is risk-driven priority setting a by-product of CERCLA and SARA requirements or is it a distinct priority-setting system?
   CERCLA compliance often leads to protocol-driven cleanups, not necessarily risk-driven cleanup. The meaning of risk-based priorities is not clear after emergency removals have eliminated known pathways.

3. How significantly do reuse-driven priorities differ from risk-driven priorities and/or CERCLA and SARA requirements? What about speed-driven or jobs-driven priorities? Can DoD's goal of harmonizing risk-driven priorities with reuse-driven priorities and CERCLA and SARA requirements be achieved?
   Under the current divisions of a base by OU and parcels, there is often a significant distinction between CERCLA compliance and reuse-driven goals. However, there is no fundamental reason this must occur and no fundamental divergence between a risk-driven approach and a reuse-driven strategy. Multiple goals can be achieved by renegotiating regulatory agreements, redrawing internal base boundaries, and focusing cleanup

efforts on the most important reuse parcels and most risky
sites. Use of removal actions can accelerate this process.

4. What is the role of project management in achieving this goal?
Do DoD policy innovations help project managers take advantage
of the flexibility existing in CERCLA and SARA?
DoD policy innovations facilitate use of flexibility existing
in CERCLA. The BRAC-cleanup-team concept of teaming DoD
project managers and regulators is particularly successful.
However, project leaders and local regulators need a better
understanding of the flexibility that exists and the interim
goals that should be achieved.

To achieve these multiple objectives for site cleanups, we
recommend the following policy ideas:

- Recognize that the total cleanup of many military bases, though
desired, is too distant and too expensive to provide realistic
policy goals or structures for project management.

- Give increased emphasis to identification of realistic interim
goals for cleanup and the obstacles to achieving them, rather
than to issues aimed at reducing long-run program costs.

- Identify and eliminate obstacles to redrawing internal base
boundaries. The USEPA should continue to move away from the
policy of fence-line-to-fence-line listing.

- Recognize that for sites remaining in federal hands pending
funds for cleanup, there is potential for contamination to
spread; this may represent a better basis for establishing
risk-based priorities than traditional exposure assessment.

- Note that despite the well-known problems of CERCLA and SARA,
it is the experience and dedication of site project management
(including regulators), and the extent of support given by
higher-level commands, that are the dominant factors in
determining failure or success. DoD's investment in human
resources for site-level management is inappropriately low
given the enormous projects being undertaken.

At a program level at DoD headquarters and/or at the USEPA, we recommend the following to policymakers:

- Provide clearer policy (as opposed to regulatory) guidance than currently exists to encourage the use of removal actions to break administrative logjams.

- Review and refine the CERCLA flexibilities in Table 1 and prepare a concise summary for project leaders and local regulators.

- Take steps to retain project leaders and regulators who have the experience to "go off-line" as they adapt projects toward achieving multiple interim policy goals. The Air Force Base Closure Authority should adopt a policy of retaining existing staff when it takes over a base rather than utilizing new BCA personnel.

- Encourage communities to develop more-strategic reuse plans that provide cleanup projects with general guidance for developing reuse-driven cleanups, even when local zoning or planning processes are still in flux.

- Recognize that the DoD project manager has a more complex obligation than simple administrative oversight of large contractors. The project manager must actively engage in project execution, be involved in all engineering decisionmaking, and not allow contractor-led projects with little policy focus to evolve. Corresponding support at the site level is required.

At the project management level, we recommend the following to DoD remedial project managers and regulators:

- Review use of the flexibilities contained in CERCLA and SARA.
- Identify interim goals by dividing each base (both active and closing) into sets of sites that are critical for reuse, for risk reduction, and for other policy objectives, and those

where cleanup can be delayed. Cleanup projects should then be focused on the most critical goals.

* Provide greater geographical focus for all environmental programs (munitions removal, lead, asbestos, historic preservation responsibilities, etc.), in addition to cleanup, for the parcels of greatest interest.

* Closely scrutinize potential economies of scale and recognize that many occur only when contemplating basewide cleanup, for which there may be insufficient funds.

* Similarly, recognize that the administrative economies of scale associated with large remedial contractors may not outweigh the advantages of a competitive contracting structure. Project leaders should tailor the contracting structure to the needs of individual bases.

We note that most of these recommendations are relevant for active bases as well as for closing bases.

MR-621-OSD

ISBN 0-8330-2327-6

9 780833 023278